FAO JECFA Monographs

ISSN 1817-7077

# COMPENDIUM OF FOOD ADDITIVE SPECIFICATIONS

Joint FAO/WHO Expert Committee on Food Additives

69th meeting 2008

FOOD AND AGRICULTURE ORGANIZATION OF THE UNITED NATIONS
Rome, 2008

The designations employed and the presentation of the material in this information
product do not imply the expression of any opinion whatsoever on the part of the Food
and Agriculture Organization of the United Nations concerning the legal or development
status of any country, territory, city or area or of its authorities, or concerning the
delimitation of its frontiers or boundaries.

ISBN 978-92-5-106065-0

All rights reserved. Reproduction and dissemination of material in this information
product for educational or other non-commercial purposes are authorized without
any prior written permission from the copyright holders provided the source is fully
acknowledged. Reproduction of material in this information product for resale or other
commercial purposes is prohibited without written permission of the copyright holders.
Applications for such permission should be addressed to:

Chief
Electronic Publishing Policy and Support Branch
Communication Division
FAO
Viale delle Terme di Caracalla, 00153 Rome, Italy
or by e-mail to:
copyright@fao.org

© FAO 2008

**SPECIAL NOTE**

While the greatest care has been exercised in the preparation of this information, FAO expressly disclaims any liability to users of these procedures for consequential damages of any kind arising out of, or connected with, their use.

# TABLE OF CONTENTS

List of participants ............................................................................................................... vii

Introduction ............................................................................................................................ ix

Specifications for certain food additives ............................................................................. 1

    Asparaginase from *Aspergillus niger* expressed in *A. niger* ................................. 3
    Calcium lignosulfonate (40-65) ............................................................................... 7
    Carob bean gum ..................................................................................................... 17
    Carob bean gum (clarified) ..................................................................................... 21
    Ethyl lauroyl arginate ............................................................................................. 25
    Guar gum ................................................................................................................ 31
    Guar gum (clarified) ............................................................................................... 35
    Iron oxides .............................................................................................................. 39
    Isomalt .................................................................................................................... 43
    Monomagnesium phosphate ................................................................................. 49
    Paprika extract Tentative ....................................................................................... 51
    Patent Blue V ......................................................................................................... 57
    Phospholipase C expressed in *Pichia pastoris* .................................................... 61
    Phytosterols, phytostanols and their esters ........................................................... 65
    Polydimethylsiloxane ............................................................................................. 71
    Steviol glycosides .................................................................................................. 75
    Sunset Yellow FCF ................................................................................................ 79
    Trisodium diphosphate .......................................................................................... 83

Withdrawal of specifications for certain food additives .................................................. 85

Analytical methods ............................................................................................................... 87

Specifications for certain flavourings ............................................................................... 89

    New specifications ................................................................................................. 91
    Spectra of certain flavourings .............................................................................. 108
    Index: Specifications of certain flavourings ....................................................... 125

Annex 1: Summary of recommendations from the 69th JECFA ................................... 127

Annex 2: Recommendations and further information required ..................................... 139

Corrigenda ........................................................................................................................... 141

# LIST OF PARTICIPANTS

## JOINT FAO/WHO EXPERT COMMITTEE ON FOOD ADDITIVES, 69$^{TH}$ MEETING
### Rome, 17 – 26 June, 2008

**Members**

Prof Jack Bend, Department of Pathology, University of Western Ontario, Canada

Prof Maria Cecilia de Figueiredo Toledo, University of Campinas, SP, Brazil

Dr Yoko Kawamura, National Institute of Health Sciences, Tokyo, Japan

Dr Paul M. Kuznesof, Silver Spring, MD, USA

Dr John C. Larsen, National Food Institute, Technical University of Denmark, Denmark (Chairman)

Dr Catherine LeClercq, National Research Institute for Food and Nutrition (INRAN), Rome, Italy

Dr Antonia Mattia, Food and Drug Administration, College Park, MD, USA

Mrs Inge Meyland, National Food Institute, Technical University of Denmark, Denmark (vice-Chairman)

Dr Gérard Pascal, INRA (Institut National de la Recherche Agronomique), L'Etang-La-Ville, France

Dr Josef Schlatter, Food Toxicology Section, Swiss Federal Office of Public Health, Zürich, Switzerland

Ms Elizabeth Vavasour, Food Directorate, Health Canada, Ottawa, Canada

Dr Madduri Veerabhadra Rao, Central Laboratories Unit, United Arab Emirates University, United Arab Emirates

Prof Ronald Walker, School of Biomedical and Molecular Sciences, University of Surrey, Surrey, UK

Mrs Harriet Wallin, National Food Safety Authority (Evira), Helsinki, Finland

Dr Brian Whitehouse, Bowdon, Cheshire, United Kingdom

**Secretariat**

Dr Peter J. Abbott, Canberra, Australia (WHO Temporary Adviser)

Ms Janis Baines, Food Standards Australia New Zealand, Canberra Australia (FAO Expert)

Dr Diane Benford, Food Standards Agency, London, United Kingdom (WHO Temporary Adviser)

Dr Annamaria Bruno, Secretariat of the Codex Alimentarius Commission, Food and Agriculture Organization, Rome, Italy (FAO Codex Secretariat)

Dr Richard Cantrill, American Oil Chemists' Society, Urbana, IL, USA (FAO Expert)

Dr Ruth Charrondiere, Nutrition and Consumer Protection Division, Food and Agriculture Organization, Rome, Italy (FAO Staff Member)

Dr Junshi Chen, Chairman of the Codex Committee on Food Additives (CCFA), Institute of Nutrition and Food Safety, Beijing, china (WHO Temporary Adviser)

Dr Myoengsin Choi, International Programme on Chemical Safety, World Health Organization, Geneva, Switzerland (WHO Staff Member)

Dr Michael DiNovi, Food and Drug Administration, College Park, MD, USA (WHO Temporary Adviser)

Dr Jean-Charles LeBlanc, French Food Safety Agency (AFSSA), Maisons Alfort, France (WHO Temporary Adviser)

Prof Symon M. Mahungu, Egerton University, Njoro, Kenya (FAO Expert)

Mrs Heidi Mattock, Tignieu Jameyzieu, France (Editor)

Dr Hyo-Min Lee, National Institute of Toxicological Research, Seoul, Republic of Korea (WHO Temporary Adviser)

Dr Ian C. Munro, CanTox Health Sciences International, Mississauga, Ontario, Canada (WHO Temporary Adviser)

Dr Utz Mueller, Food Standards Australia New Zealand, Canberra Australia (WHO Temporary Adviser)

Dr Zofia Olempska-Beer, US Food and Drug Administration, College Park, MD, USA (FAO Expert)

Mrs Marja E.J. Pronk, Center for Substances and Integrated Risk Assessment, National Institute for Public Health and the Environment, Bilthoven, the Netherlands (WHO Temporary Adviser)

Prof Andrew G. Renwick, Clinical Pharmacology Group, University of Southampton, Southampton, United Kingdom (WHO Temporary Adviser)

Prof I. Glenn Sipes, Department of Pharmacology, College of Medicine, University of Arizona, Tucson, AZ, USA (WHO Temporary Adviser)

Dr Klaus Schneider, FoBiG (Forschungs- und Beratungsinstitut Gefahrstoffe GmbH), Freiburg, Germany (WHO Temporary Adviser)

Dr Angelika Tritscher, International Programme on Chemical Safety, World Health Organization, Geneva, Switzerland (WHO Joint Secretary)

Dr Takashi Umemura, National Institute of Health Sciences, Tokyo, Japan (WHO Temporary Adviser)

Dr Annika Wennberg, Nutrition and Consumer Protection Division, Food and Agriculture Organization, Rome, Italy (FAO Joint Secretary)

Prof Gary M Williams, Environmental Pathology and Toxicology, New York Medical College, Valhalla, NY, USA (WHO Temporary Adviser)

# INTRODUCTION

This volume of FAO JECFA Monographs contains specifications of identity and purity prepared at the 69th meeting of the Joint FAO/WHO Expert Committee on Food Additives (JECFA), held in Rome on 17-26 June 2008. In addition, a revised analytical method of assay of nickel in polyols was prepared and included in this publication. The specifications monographs are one of the outputs of JECFA's risk assessment of food additives, and should be read in conjunction with the safety evaluation, reference to which is made in the section at the head of each specifications monograph. Further information on the meeting discussions can be found in the summary report of the meeting (see Annex 1), and in the full report which will be published in the WHO Technical Report series. Toxicological monographs of the substances considered at the meeting will be published in the WHO Food Additive Series.

Specifications monographs prepared by JECFA up to the 65th meeting, other than specifications for flavouring agents, have been published in consolidated form in the Combined Compendium of Food Additive Specifications which is the first publication in the series FAO JECFA Monographs. This publication consist of four volumes, the first three of which contain the specifications monographs on the identity and purity of the food additives and the fourth volume contains the analytical methods, test procedures and laboratory solutions required and referenced in the specifications monographs. FAO maintains an on-line searchable database of all JECFA specifications monographs from the FAO JECFA Monographs, which is available at: http://www.fao.org/ag/agn/jecfa-additives/search.html . The specifications for flavourings evaluated by JECFA, and previously published in FAO Food and Nutrition Paper 52 and subsequent Addenda, are included in a database for flavourings (flavouring agent) specifications which has been updated and modernized. All specifications for flavourings that have been evaluated by JECFA since its 44th meeting, including the 69th meeting, are available in the new format online searchable database at the JECFA website at FAO: http://www.fao.org/ag/agn/jecfa-flav/search.html. The databases have query pages and background information in English, French, Spanish, Arabic and Chinese. Information about analytical methods referred to in the specifications is available in the Combined Compendium of Food Additive Specifications (Volume 4), which can be accessed from the query pages.

An account of the purpose and function of specifications of identity and purity, the role of JECFA specifications in the Codex system, the link between specifications and methods of analysis, and the format of specifications, are set out in the Introduction to the Combined Compendium, which is available in shortened format online on the query page, which could be consulted for further information on the role of specifications in the risk assessment of additives.

Chemical and Technical Assessments (CTAs) for some of the food additives have been prepared as background documentation for the meeting. These documents are available online at: http://www.fao.org/ag/agn/agns/jecfa_archive_cta_en.asp .

*Contact and Feedback*

More information on the work of the Committee is available from the FAO homepage of JECFA at: http://www.fao.org/ag/agn/agns/jecfa_index_en.asp . Readers are invited to address comments and questions on this publication and other topics related to the work of JECFA to:

jecfa@fao.org

# SPECIFICATIONS FOR CERTAIN FOOD ADDITIVES

## *New and revised specifications*

New (N) or revised (R) specifications monographs were prepared for the following food additives and these are provided in this publication:

Asparaginase from *Aspergillus niger* expressed in *A. niger* (N)
Calcium lignosulfonate (40-65) (N)
Carob bean gum (R)
Carob bean gum (clarified) (R)
Ethyl lauroyl arginate (N)
Guar gum (R)
Guar gum (clarified) (R)
Iron oxides (R)
Isomalt (R)
Monomagnesium phosphate (N)
Paprika extract (N) Tentative
Patent Blue V (R)
Phospholipase C expressed in *Pichia pastoris* (N)
Phytosterols, phytostanols and their esters (N)
Polydimethylsiloxane (R)
Sunset Yellow FCF (R)
Steviol glycosides (R)
Trisodium diphosphate (N)

In the specifications monographs that have been assigned a tentative status, there is information on the outstanding information and a timeline by which this information should be submitted to the FAO JECFA Secretariat.

In addition to these specifications monographs, minor revisions were made to the specifications monographs for the food additives Canthaxanthin, Chlorophyllins, copper complexes sodium and potassium salts and Fast Green FCF. The Committee decided that republication in the FAO JECFA Monongraphs of these specifications monographs were not necessary.

Canthaxanthin: The Committee was made aware that in the specifications for canthaxanthin, the wording of the criterion for the assay could be misinterpreted. The Committee decided to change the original text "Not less than 96% of total colouring matters (expressed as canthaxanthin)" in the electronic version of the specifications on the FAO JECFA website to read: "Not less than 96% total colouring matters (expressed as canthaxanthin)."

Chlorophyllins, copper complexes sodium and potassium salts: The Committee was informed that the Colour Index (C.I.) International number in the specifications for chlorophyllin, copper complexes sodium and potassium salts was incorrectly stated. The Committee decided to include the correct number, C.I. No. 75815, in the electronic version of the specifications on the FAO JECFA website.

Fast Green FCF: The Committee was informed that an error had been introduced into the specification for Fast Green FCF published in the Combined Compendium of Food Additive Specifications (2005) when the text from FAO Food and Nutrition Paper 52 was transcribed. The value for absorptivity in the determination of the quantity of leuco base was corrected to read 0.156 in the electronic version of the specifications on the FAO JECFA website.

New and revised INS numbers assigned to food additives by the Codex Alimentarius Commission at its 31st session in 2008, (ALINORM 08/31/12, Appendix XII) have been introduced in the corresponding JECFA food additive specifications monographs in the on-line database, as appropriate, and these are not reproduced in this publication.

The chemical abstract numbers (C.A.S.) for the food additive Dicalcium pyrophosphate has been revised to 7790-76-3 in the specifications monographs in the on-line database and is not reproduced in this publication.

## ASPARAGINASE from *ASPERGILLUS NIGER* expressed in *A. NIGER*

*New specifications prepared at the 69th JECFA (2008), published in FAO JECFA Monographs 5 (2008). An ADI "not specified" was established at the 69th JECFA (2008).*

| | |
|---|---|
| **SYNONYMS** | Asparaginase II; L-asparaginase; α-asparaginase |
| **SOURCES** | Asparaginase is produced by submerged fed-batch fermentation of a genetically modified strain of *Aspergillus niger* which contains the asparaginase gene derived from *A. niger*. The enzyme is isolated from the fermentation broth by filtration to remove the biomass and concentrated by ultrafiltration. The enzyme concentrate is subjected to germ filtration and is subsequently formulated and standardized to the desired activity using food-grade compounds. |
| Active principles | Asparaginase |
| Systematic names and numbers | L-Asparagine amidohydrolase; EC 3.5.1.1; CAS No. 9015-68-3 |
| Reactions catalysed | Hydrolysis of L-asparagine to L-aspartic acid and ammonia |
| Secondary enzyme activities | No significant levels of secondary enzyme activities. |
| **DESCRIPTION** | Yellow to brown clear liquid or off-white granulates |
| **FUNCTIONAL USES** | Enzyme preparation.<br>Used in food processing to reduce the formation of acrylamide from asparagine and reducing sugars during baking or frying. |
| **GENERAL SPECIFICATIONS** | Must conform to the latest edition of the JECFA General Specifications and Considerations for Enzyme Preparations Used in Food Processing. |

## CHARACTERISTICS

IDENTIFICATION

<u>Asparaginase activity</u>   The sample shows asparaginase activity.
See description under TESTS.

## TESTS

<u>Asparaginase activity</u>

**Principle**
Asparaginase catalyses the conversion of L-asparagine to L-aspartic acid and ammonia. The liberated ammonia subsequently reacts with phenol nitroprusside and alkaline hypochlorite resulting in a blue colour (known as Berthelot reaction). The activity of asparaginase is determined by measuring the absorbance of the reaction mixture at 600 nm.

The asparaginase activity is expressed in ASPU units. One ASPU is defined as the amount of the enzyme required to liberate one micromole of ammonia from L-asparagine per minute under the conditions of the assay (pH=5.0; 37°).

Note: The measuring range of the method is 1.5 – 12 ASPU/ml.

**Apparatus**
Spectrophotometer (600 nm)
Water bath with thermostatic control (37±0.1°)
pH meter
Vortex mixer
Magnetic stirrer
Disposable culture tubes (glass, 10x100 mm)

**Reagents and solutions**
(Note: use Ultra High Quality water with conductivity of ≤ 0.10 µS/cm)

*Phenol nitroprusside solution* (Sigma-Aldrich P6994 or equivalent)

*Sodium hypochlorite 0.2% in alkali solution* (Sigma-Aldrich A1727 or equivalent)

*Sodium hydroxide solution 4 M*: Weigh 160 g of NaOH pellets. Dissolve in approximately 800 ml of water in a 1 l volumetric flask. Cool down to room temperature, add water to volume and mix until fully dissolved. The solution is stable for 3 months at room temperature.

*Citric acid dilution buffer 0.1M, pH 5.00±0.03*: Weigh 21.01 g of citric acid monohydrate (analytical reagent grade). Dissolve in approximately 900 ml of water in a 1 l volumetric flask. Adjust the pH to 5.00±0.03 with 4 M NaOH. Add water to volume and mix. The solution is stable for 1 month when stored in a refrigerator.

*L-asparagine substrate solution:* Weigh 1.50 g of L-asparagine (L-asparagine monohydrate ≥ 99%, Sigma-Aldrich A8381 or equivalent). Dissolve in approximately 80 ml of the citric acid dilution buffer in a 100 ml volumetric flask and stir on a magnetic stirrer until completely dissolved. Add the dilution buffer to volume and mix. The solution should be freshly prepared before the analysis.

*TCA stop solution:* Weigh 25 g of trichloroacetic acid (Sigma-Aldrich 27242 (Riedel-de Haen) or equivalent). Dissolve in approximately 80 ml of water in a 100 ml volumetric flask. Add water to volume and mix. The solution is stable for 1 year at room temperature.

*Standard solution*: Weigh to ± 0.1 mg approximately 3.9 g of ammonium sulfate (analytical reagent grade) with an officially certified content. Dissolve in approximately 40 ml of the citric acid dilution buffer in a 50 ml volumetric flask by stirring on a magnetic stirrer for about 15 min. Add the dilution buffer to volume and mix. Make five dilutions with the dilution buffer and calculate the concentration of each dilution based on the certified content of ammonium sulfate. The table below provides an example.

| Label | Dilution factor | Concentration, mg/ml |
|-------|----------------|----------------------|
| S1    | 60             | 1.3                  |
| S2    | 30             | 2.6                  |
| S3    | 10             | 7.8                  |
| S4    | 6              | 13.0                 |
| S5    | 4              | 19.5                 |

*Control sample solution*: Weigh to ± 0.1 mg an amount of an asparaginase preparation with known activity (for example, 18930 ASPU/g; batch KFP0445A/DIV/4; expiration date January 2020; available from DSM Food Specialties) approximately equivalent to 4000 ASPU. Dissolve in approximately 20 ml of the citric acid dilution buffer in a 25 ml volumetric flask. Add the dilution buffer to volume, and mix. Dilute the solution with the dilution buffer to a final activity of approximately 6 ASPU/ml.

*Test sample solution*: Weigh to ± 0.1 mg approximately 2.5 g of an asparaginase preparation. Dissolve in approximately 20 ml of the citric acid dilution buffer in a 25 ml volumetric flask. Add the dilution buffer to volume and mix. Dilute the solution with the dilution buffer to a final activity of approximately 6 ASPU/ml.

**Procedure**
*Standard curve:*

1. Label five test tubes according to the concentrations of the standard solutions (S1 to S5). Pipette 2.0 ml of the substrate solution to each tube. Incubate in the water bath for 10 minutes. To each tube, add 100 µl of the appropriate standard solution and mix. Incubate the tubes in the water bath exactly for 30 min. Add 0.4 ml of the TCA stop solution to stop the reaction. Add 2.5 ml of water and mix. This is the reaction mixture.

2. Prepare five test tubes (labeled S1 to S5). Add to each tube 800 µl of water and 20 µl of the appropriate reaction mixture. To develop colour, add 170 µl of the phenol nitroprusside solution, mix and add 170 µl of the alkaline sodium hypochlorite solution. Mix and incubate in the water bath for 10 min. Transfer the content of each tube to the spectrophotometer cuvette and measure the absorbance at 600 nm after zeroing the instrument against air.

3. Use linear regression to prepare the standard curve. Plot the absorbance against the concentration of ammonium sulfate in the standard solutions (mg/ml). Use the slope of the standard curve (ml/mg) to calculate the activity of the control and test samples.

(NOTE: The standard curve should be prepared immediately prior to sample analysis.)

*Control and test samples:*

1. For all control and test samples, follow the procedure described in steps 1 and 2 above for the standard solutions.

2. Use a blank for each control and test sample. To prepare the blank, pipette into a test tube 2.0 ml of the substrate solution and 0.4 ml of the TCA stop reagent. Mix and add 100 µl of either the control or test sample solution. Mix and incubate in the water bath for 30 min. Add 2.5 ml of water and continue as described in step 2 of the procedure for the standard solutions.

**Calculations**

Calculate the activity of each control and test sample in activity units per gram of the enzyme preparation (ASPU/g) using the following formula:

$$ASPU/g = \frac{A \times V \times Df \times 2 \times 10^6}{a \times M \times W \times 30 \times 10^3}$$

Where:

A is the absorbance of the sample minus the absorbance of the blank

V is the initial volume of the sample solution (25 ml)

Df is the dilution factor

2 accounts for 2 moles of ammonia per 1 mole of ammonium sulfate

$10^6$ is the conversion factor from moles to µmoles

a is the slope of the standard curve (ml/mg)

M is the molar mass of ammonium sulfate (132.14 g/mol)

W is the sample weight (g)

30 is the reaction time (min)

$10^3$ is the conversion factor from milligrams to grams

# CALCIUM LIGNOSULFONATE (40-65)

*New specifications prepared at the 69th JECFA (2008), published in FAO JECFA Monographs 5 (2008). An ADI of 0-20 mg/kg bw was established at the 69th JECFA (2008).*

| | |
|---|---|
| **SYNONYMS** | Lignosulfonic acid, calcium salt (40-65) |
| **DEFINITION** | Calcium lignosulfonate (40-65) is an amorphous material obtained from the sulfite pulping of softwood. The lignin framework is a sulfonated random polymer of three aromatic alcohols: coniferyl alcohol, *p*-coumaryl alcohol, and sinapyl alcohol, of which coniferyl alcohol is the principle unit. After completion of the pulping, the water-soluble calcium lignosulfonate is separated from the cellulose, purified (ultrafiltration), and acidified. The recovered material is evaporated and spray dried. The commercial product has a weight-average molecular weight range of 40,000 to 65,000. |
| **DESCRIPTION** | Light yellow-brown to brown powder |
| **FUNCTIONAL USES** | Carrier |

**CHARACTERISTICS**

IDENTIFICATION

| | |
|---|---|
| Solubility (Vol. 4) | Soluble in water. Practically insoluble in organic solvents. |
| IR spectrum (Vol. 4) | The infrared absorption spectrum of a potassium bromide pellet of dried sample exhibits characteristic absorptions at 1210-1220 $cm^{-1}$, 1037 $cm^{-1}$, and 655 $cm^{-1}$. |
| UV spectrum (Vol. 4) | A 0.05% sample solution is diluted 1:10 and adjusted to a pH of 2.0-2.2 by addition of 3 drops of 5 M hydrochloric acid. This solution exhibits an absorption maximum at 280 nm. |
| Weight-average molecular weight | Between 40,000 to 65,000 (>90% of the sample ranges from 1,000 to 250,000)<br>See description under TESTS |
| pH (Vol. 4) | 2.7 - 3.3 (10% solution) |
| Calcium (Vol. 4) | Passes test ("General Methods, Identification Tests," Volume 4) |
| Degree of sulfonation | 0.3 – 0.7, on the dried basis<br>See description under TESTS |

PURITY

| | |
|---|---|
| Calcium | Not more than 5.0 %, on the dried basis<br>See description under TESTS |
| Loss on drying (Vol. 4) | Not more than 8.0% (105°, 24 h) |

| | |
|---|---|
| Reducing sugars | Not more than 5.0%, on the dried basis<br>See description under TESTS |
| Sulfite | Not more than 0.5%, on the dried basis<br>See description under TESTS |
| Total Ash | Not more than 14.0%, on the dried basis<br>See description under TESTS |
| Arsenic (Vol. 4) | Not more than 1 mg/kg<br>Determine by the atomic absorption hydride technique. The selection of sample size and method of sample preparation may be based on the principles of the methods described in Volume 4 (under "General Methods, Metallic Impurities"). Alternatively, determine arsenic using Method II of the Arsenic Limit Test, taking 3 g of the sample rather than 1 g, following the procedure for organic compounds. |
| Lead (Vol. 4) | Not more than 2 mg/kg<br>Determine using an AAS/ICP-AES technique appropriate to the specified level. The selection of sample size and method of sample preparation may be based on the principles of the methods described in Volume 4 (under "General Methods, Metallic Impurities"). |

## TESTS

IDENTIFICATION TESTS

Weight-average molecular weight

Principle
Size-exclusion chromatography is used to obtain the molecular-weight distribution profile of the sample.

Reagents
(NOTE: All solutions and dilutions are to be made using distilled, deionized water)
Dimethylsulfoxide (DMSO), HPLC grade.
Disodium hydrogen phosphate ($Na_2HPO_4 \cdot 7H_2O$), Reagent grade
50 % sodium hydroxide (NaOH), Reagent grade
Sodium dodecylsulfate (SDS), Gradient grade (ultra grade)

Equipment
Size-exclusion chromatograph (Agilent Technologies or equivalent) equipped with autosampler, HPLC-pump, degassing unit, UV-detector or RI-detector, MALLS (Multi-Angle Laser Light Scattering) detector (Wyatt Technology or equivalent) with interference filters.
Columns - Glucose-divinylbenzene (DVB), $10^4$ Å pore size, 500x10 mm (Jordi Associates or equivalent ) and TSK gel PWXL 6 mm x 4 cm guard column (TOSOH Bioscience or equivalent)
Syringe filter - 0.2 µm GHP (Pall Corp. or equivalent)
Filter paper - 0.22 µm Millipore GSWP (Millipore Corp. or equivalent)

Eluent
Weigh 1600.0 g of water into a 2 litre flask. Add 161.8 g DMSO, mix, and add 21.44 g $Na_2HPO_4 \cdot 7H_2O$. Adjust the pH to 10.5 with NaOH, add 1.6 g of SDS, and filter the mixture through the GSWP filter paper.

### Sample solution
Accurately weigh and transfer 20 mg of previously dried sample into a 10-ml volumetric flask and dilute to the mark with water. Using the syringe filter, filter the solution into a vial.

### Procedure
Set the oven temperature of the chromatograph at 60°. Begin the flow of eluent (1.0 ml/min - the pressure should not exceed 1000 psi.) through the chromatography system. After at least one hour has elapsed, inject the Sample solution (20 µl) onto the column and record the chromatograph. Calculate the weight-average molecular weight from the chromatogram using suitably certified software.

## Degree of sulfonation

### Principle
The Degree of sulfonation is the ratio of Organic sulfur to the Methoxyl content of the sample. Organic sulfur is calculated as the difference between Total sulfur (determined by elemental analysis) and Inorganic sulfur (determined by ion chromatography).

**Determination of Total sulfur**

### Equipment and Reagents
Elemental Analyser (Thermo Fisher Scientific or equivalent)
Analytical balance
Tin capsules
BBOT standard (2,5-(Bis(5-tert-butyl-2-benzo-oxazol-2-yl) thiophene))
Vanadium pentoxide

### Analytical conditions
| | |
|---|---|
| Carrier gas - Helium | 120 ml/min |
| Combustion furnace temp. | 1000° |
| Oven temp. | 70° |
| Helium pressure | 150 kPa |
| Oxygen pressure | 150 kPa |
| Oxygen loop | 5 ml |
| Run time | 300 sec. |

### System checks
Vanadium pentoxide
Vanadium pentoxide and BBOT

### Procedure
*System checks*: Introduce small amounts of the two System checks separately into two tin capsules (no need to weigh). Run the two System checks through the analyzer. Observation of a sulfur peak in the chromatogram confirms that the system is working properly.

*Standards*: Introduce approximately 0.2 mg of vanadium pentoxide into each of four tin capsules and weigh them. Accurately weigh 0.5, 1.0, 1.5 and 2.0 mg of BBOT standard into the four capsules. Run the four standards through the analyzer and construct a calibration curve. The calibration curve should be a straight line with a correlation coefficient of at least 0.999.

*Sample*: Introduce approximately 0.2 mg of vanadium pentoxide into each of two tin capsules and weigh them. Accurately weigh 1-2 mg of sample, previously dried, into each capsule and run them through the analyzer. Run additional samples in duplicate. After every fourth sample, accurately weigh 0.5-2.0 mg of the BBOT standard into a tared tin capsule containg 0.2 mg of vanadium pentoxide to run as a control. (NOTE: The weight of BBOT taken is chosen to fall within the calibration curve.) The standard deviation of the control BBOT standard should be no more than 0.20. Obtain the weight (mg) of total sulfur for each sample (w) from the calibration curve and calculate the percent Total sulfur for each by dividing by the weight of the corresponding sample taken (W) using the formula:

$$\% \text{ Total sulfur} = 100 \times w/W$$

Compute the average % Total sulfur.

### Determination of Inorganic sulfur
(NOTE: All solutions and dilutions to be made using distilled, deionized water)

Equipment
Ion Chromatograph (Dionex Corporation or equivalent) with conductivity detector and autosampler
Anion Self-Regenerating Suppressor (ASRS-Ultra 4 or equivalent)
Analytical Column - IonPac AS 11 (Dionex Corporation or equivalent)
Guard Column - IonPac AG 11 (Dionex Corporation or equivalent)
Syringe filter - 0.2 $\mu$m GHP (Pall Corp. or equivalent)

Reagents
0.1 M NaOH (sodium hydroxide): 5.265 ml 50% NaOH (Reagent grade), diluted to 1000 ml
1% NaOH (sodium hydroxide): 2 ml 50% NaOH (Reagent grade), diluted to 100 ml
3% $H_2O_2$ (hydrogen peroxide): 50 ml 30% $H_2O_2$ (Reagent grade), diluted to 500 ml
Eluent: 0.1 M NaOH/water (10/90)

Stock standard solution
1 mg/ml, prepared by dissolving 0.1479 g sodium sulfate in 100 ml of water

Standard sulfate solutions (2.0 mg/l, 5.0 mg/l, 20.0 mg/l, and 40.0 mg/l)
Pipet 0.1, 0.25, 1.0 and 2.0 ml of the Stock standard solution into separate 50-ml volumetric flasks. Add 1 ml of 3 % $H_2O_2$, dilute to volume with water, and mix.

Sample solution
Accurately weigh and transfer 30 mg of previously dried sample into a 50-ml volumetric flask and dissolve it in 10 ml of 1% NaOH. Add 5 ml of 3% $H_2O_2$ and allow to stand overnight. Then, dilute to volume with water.

Procedure
(NOTE: Filter all solutions through the syringe filter prior to injection into the ion chromatograph.) Set the eluant flow rate to 0.7 ml/min.

Separately inject 10 μl of the standard sulfate solutions and the Sample solution and record the chromatograms for a run time of 15 min. (NOTE: The sulfate retention time is 7 min.) Construct a calibration curve and determine the sulfate concentration of the Sample solution. Determine the weight (mg) of sulfate in the sample, w, and calculate the percentage of Inorganic sulfur in the sample using the following equation:

$$\% \text{ Inorganic sulfur} = 100 \times w \times 32/(W \times 96)$$

where
- W is the weight (mg) of the sample taken
- 32 is the formula weight of sulfur
- 96 is the formula weight of sulfate

**Determination of Organic sulfur**

$$\% \text{ Organic sulfur} = (\% \text{ Total sulfur}) - (\% \text{ Inorganic sulfur})$$

**Determination of Methoxyl (-OCH$_3$)**

Principle
Heating with hydroiodic acid decomposes the sample to form methyl iodide which reacts to form iodine. The iodine is quantitatively determined by titration with sodium thiosulfate.

Reagents
Phenol, Reagent grade
Hydroiodic acid, HI, (min. 57%), Reagent grade
Red phosphorus
5% Cadmium sulfate (CdSO$_4$) solution
Bromine, Reagent grade
Formic acid (concentrated), Reagent Grade
1 M Sulfuric acid (H$_2$SO$_4$), Reagent grade
10% Potassium iodide solution (KI), Reagent grade
0.025 M Sodium thiosulfate (Na$_2$S$_2$O$_3$), Reagent grade
Acetic acid (glacial) saturated with Sodium acetate, Reagent grade
3 % Sodium carbonate (Na$_2$CO$_3$) solution

Equipment (Anal. Chem. Acta, vol. 15 (1956) p. 279-283)

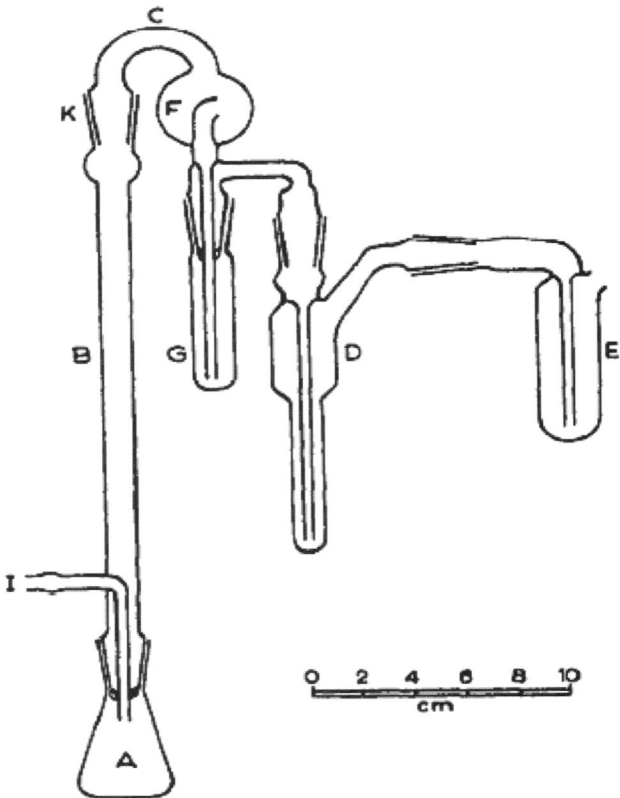

## Procedure

Accurately weigh 15-20 mg of previously dried sample on a small square of aluminium foil. Wrap the foil around the sample and put it into the reaction flask (A) to which 5 ml of hydroiodic acid, approx. 2 g of phenol, and a few glass beads have been added. Add 5 ml of 5% cadmium sulfate solution containing about 0.3 mg of red phosphorus into the washer (G). Add 10 ml of acetic acid (saturated with sodium acetate) and 10 droplets of bromine to the receiver (D). Finally, fill the U-trap (E) with sodium hydroxide or other suitable absorbant that will prevent bromine from leaving the system.

Pass nitrogen gas through a 3% $Na_2CO_3$ solution and into the system through the side arm (I) of the air condensor (B). Heat the reaction flask (A) to 140-145° for 1 hour in a glycerin bath. Wash the contents of the receiver (D) into a 250 ml Erlenmeyer flask containing 10 ml of acetic acid (saturated with sodium acetate). Rotate the flask and add formic acid dropwise until the colour disappears. Add 5 ml 10 % potassium iodide solution and mix. Then add 10 ml of 1 M sulfuric acid and let the flask stand for 3 minutes. Titrate the solution with 0.025 M $Na_2S_2O_3$ until the colour changes from yellowish to colourless. Calculate the percent methoxyl from the following equation:

$$\% \text{ Methoxyl} = V \times 0.025 \times 31 \times 100/(W \times 6 \times 1000)$$

where
   V is the volume (ml) of sodium thiosulfate used in the titration
   W is the weight (mg) of the sample taken
   0.025 is the concentration of the sodium thiosulfate
   31 is the formula weight of methoxyl
   6 is stoichiometric conversion factor between the titrant and the methoxyl moiety

**Degree of sulfonation**
Calculation

(% Organic sulfur)/(% methoxyl)

PURITY TESTS

Calcium

Reagents
(NOTE: All solutions and dilutions to be made using distilled, deionized water)
Calcium reference standard, Certified 1000 ppm (Mallinckrodt or equivalent)
Nitric acid (65%), Reagent grade
Hydrogen peroxide (30%), Reagent grade
Cesium chloride, suprapur
Ionization buffer: 12.1 mg/ml of cesium chloride

Standard calcium solution
3 µg/ml, prepared by diluting with water 1.5 ml of the Calcium reference standard to 500 ml. Store in polyethylene bottles.

Sample solution
Accurately weigh 0.2 g of a previously dried sample into a graduated Pyrex flask. Add 5 ml of 65% nitric acid and 2 ml of 30% hydrogen peroxide. Boil the sample for 1 hour in a microwave oven. Dilute the sample stepwise and quantitatively to a suitable concentration level with purified water (< 0.00007 mS). A sample with 5% Calcium should be diluted by a factor of 5000 to give a final concentration of 2 µg/ml.

Procedure
Using a suitable atomic absorbtion spectrophotometer optimized according to the manufacturer's instructions, measure the absorbance of the Sample solution at 422.7 nm. By dilution of the working standard (manually or using the auto-diluter of the instrument) prepare solutions for constructing a 4-point calibration curve to correspond to a calcium content in the range 0 – 7.5 %, The sample and standard solutions and the Ionization buffer are mixed automatically by the sampling system of the instrument. Set the mixing ratio for standard/sample solutions to Ionization buffer at 3:1. Obtain the calcium concentration of the Sample solution from the calibration curve, determine the weight (g) of calcium in the sample, w, and calculate the percent of calcium in the previously dried sample from the equation:

$$\% \text{ Calcium} = 100 \times w/W$$

where W is the weight (g) of sample taken.

Reducing sugars

Principle
Reducing sugars react with p-hydroxybenzoichydrazide (PHBH) in alkaline environments. The substance formed absorbs yellow light at 410 nm. Calcium is used to enhance the colour.

Equipment
Flow Injection Analyser (O.I. Analytical or equivalent)
Cellulose membranes, Type C 25 MM (Astoria-Pacific or equivalent)

### Reagents
Glucose, anhydrous quality for biochemistry analysis
Brij-35 ((Polyoxyethyleneglycol dodecyl ether), ultra grade (O.I. Analytical or equivalent)
Calcium Chloride, $CaCl_2$, Reagent grade
Citric Acid, Reagent grade
Hydrochloric Acid, HCl, Reagent grade
1 M Sodium Hydroxide, NaOH, Reagent grade
PHBH, p-Hydroxybenzoichydrazide (Sigma-Aldrich or equivalent)

### Standard glucose solutions
100 mg/l, 1000 mg/l, and 2000 mg/l, prepared using deionized water

### Sample solution
Accurately weigh 0.5 g of a previously dried sample into a 50-ml volumetric flask. Dissolve and dilute to volume with deionized water.

### Procedure
(NOTE: Set the analyzer flow to the "low" position on both pumps and the temperature of the heater to 90°. The instrument should stabilize in about 15 minutes. The signal should be less than ±1000 micro-Absorbance Units before starting the analysis.) Introduce separately 100 µl of each of the Sample solution and Standard glucose solutions into the analyzer. For each analysis, air is introduced followed by addition of 0.2% Brij-35 at a continuous flow of 0.287 ml/min. The solutions are then dialyzed through a cellulose membrane. After dialysis, add 1M NaOH at 0.385 ml/min, $CaCl_2$ and PHBH, both at 0.074 ml/min, into the mixing chamber of the analyzer. The mixture then enters the heater (previously set at 90°) where bubbles are eliminated, after which it reaches the detector (set at 410 nm).

Run duplicate injections of every Sample solution. Construct a calibration curve from the Standard glucose solutions and obtain the concentration of reducing sugars in the Sample solution. Determine the weight (mg) of reducing sugars in the sample, w, and calculate the percentage of reducing sugars in the sample using the equation:

$$\% \text{ Reducing sugars} = 100 \times w/W$$

where
    W is the weight (mg) of sample taken

## Sulfite

### Principle
Sulfite is stabilized in an aqueous solution with formaldehyde and subsequently separated from other anions utilizing an ion-exchange column.

### Equipment
Ion Chromatograph ((Dionex Corporation or equivalent) with conductivity detector and autosampler
Anion Self-Regenerating Suppressor (ASRS-Ultra 4 or equivalent)
Analytical Column - IonPac AS 11 (Dionex Corporation or equivalent)
Guard Column - IonPac AG 11 (Dionex Corporation or equivalent)
Syringe filter - 0.2 µm GHP (Pall Corp.or equivalent)

### Reagents
(NOTE: All solutions and dilutions to be made using distilled, deionized water.)

Formaldehyde (37%), Reagent grade

Formaldehyde solution: 0.5 ml Formaldehyde (37%) diluted to 1000 ml (Prepare fresh on day of use.)

Sodium Sulfite ($Na_2SO_3$), Reagent grade.

0.1 M Sodium Hydroxide (NaOH), Reagent grade

### Eluant
0.1 M NaOH/water (10/90)

### Stock standard solution
1 mg/ml, prepared with 0.1574 g $Na_2SO_3$ in 100 ml of Formaldehyde solution.

### Standard sulfite solutions
2.0 mg/l, 5.0 mg/l, 10.0 mg/l, and 20.0 mg/l, made with freshly prepared Formaldehyde solution

### Sample solution
Accurately weigh and transfer about 0.15 g of sample, previously dried, into a 50-ml volumetric flask. Dilute to mark with Formaldehyde solution.

### Procedure
(NOTE: Filter all solutions before injection into the Ion Chromatograph.) The chromatographic system is run isocratically with eluent flow rate of 0.7 ml/min. Separately inject 10 µl of the Standard sulfite solutions and the Sample solution and record the chromatograms for a run time of 15 min. The sulfite retention time is 6 min. Construct a calibration curve and determine the sulfite concentration of the Sample solution. Determine the weight (mg) of sulfite in the sample, w, and calculate the percentage of sulfite in the sample using the following equation:

$$\% \text{ Sulfite} = 100 \times w/W$$

where W is the weight (mg) of sample taken.

## Total Ash

Accurately weigh 0.5 -1 g of a previously dried sample in a tared platinum crucible that has been cleaned with potassium bisulfate and dried at 105°. Heat the sample cautiously over a flame. Ignite at 550° for 1 hour, and then at 900° for at least 10 minutes, until all dark particles have disappeared and the ash is white. Allow the ash to cool in a desiccator and determine the weight (mg) of the residue ($W_R$).

$$\% \text{ Ash} = 100 \times W_R/W_S$$

where $W_S$ (mg) is the weight of sample taken.

# CAROB BEAN GUM

*Prepared at the 69th JECFA (2008), published in FAO JECFA Monographs 5 (2008), superseding tentative specifications prepared at the 67th JECFA (2006) and published in FAO JECFA Monographs 3 (2006). An ADI "not specified" was established at the 25th JECFA (1981).*

| | |
|---|---|
| **SYNONYMS** | Locust bean gum, INS No. 410 |
| **DEFINITION** | Primarily the ground endosperm of the seeds from *Ceratonia siliqua* (L.) Taub. (Fam. *Leguminosae*) mainly consisting of high molecular weight (approximately 50,000-3,000,000) polysaccharides composed of galactomannans; the mannose:galactose ratio is about 4:1. The seeds are dehusked by treating the kernels with dilute sulfuric acid or with thermal mechanical treatments, elimination of the germ followed by milling and screening of the endosperm to obtain native carob bean gum. The gum may be washed with ethanol or isopropanol to control the microbiological load (washed carob bean gum). |
| C.A.S. number | 9000-40-2 |
| Structural formula | |
| **DESCRIPTION** | White to yellowish white, nearly odourless powder |
| **FUNCTIONAL USES** | Thickener, stabilizer, emulsifier, gelling agent |
| **CHARACTERISTICS** | |
| IDENTIFICATION | |
| Solubility (Vol. 4) | Insoluble in ethanol |
| Gel formation | Add small amounts of sodium borate TS to an aqueous dispersion of the sample; a gel is formed. |
| Viscosity | Transfer 2 g of the sample into a 400-ml beaker and moisten thoroughly with about 4 ml of isopropanol. Add 200 ml of water with vigorous stirring until the gum is completely and uniformly dispersed. |

An opalescent, slightly viscous solution is formed. Transfer 100 ml of this solution into another 400-ml beaker. Heat the mixture in a boiling water bath for about 10 min and cool to room temperature. There is an appreciable increase in viscosity (differentiating carob bean gums from guar gums).

Gum constituents (Vol. 4)   Proceed as directed under Gum Constituents Identification using 100 mg of the sample instead of 200 mg and 1 to 10 µl of the hydrolysate instead of 1 to 5 µl. Use galactose and mannose as reference standards. These constituents should be present.

Microscopic examination   Disperse a sample of the gum in an aqueous solution containing 0.5% iodine and 1% potassium iodide on a glass slide and examine under a microscope. Carob bean gum contains long stretched tubiform cells, separated or slightly interspaced. Their brown contents are much less regularly formed than in Guar gum.

PURITY

Loss on drying (Vol. 4)   Not more than 14% (105°, 5 h)

Total ash (Vol. 4)   Not more than 1.2% (800°, 3-4 h)

Acid-insoluble matter (Vol. 4)   Not more than 4.0%

Protein (Vol. 4)   Not more than 7.0%
Proceed as directed under Nitrogen Determination (Kjeldahl Method) in Volume 4 (under "General Methods, Inorganic components"). The percentage of nitrogen determined multiplied by 6.25 gives the percentage of protein in the sample.

Starch   To a 1 in 10 dispersion of the sample add a few drops of iodine TS; no blue colour is produced.

Residual solvents   Not more than 1% of ethanol or isopropanol, singly or in combination
See description under TESTS

Lead (Vol. 4)   Not more than 2 mg/kg
Determine using an AAS/ICP-AES technique appropriate to the specified level. The selection of sample size and method of sample preparation may be based on the principles of the methods described in Volume 4 (under "General Methods, Metallic Impurities").

Microbiological criteria (Vol. 4)   Initially prepare a $10^{-1}$ dilution by adding a 50 g sample to 450 ml of Butterfield's phosphate-buffered dilution water and homogenizing the mixture in a high-speed blender.

Total (aerobic) plate count: Not more than 5,000 CFU/g
*E. coli:* Negative in 1g
*Salmonella:* Negative in 25 g
Yeasts and moulds: Not more than 500 CFU/g

# TESTS

PURITY TESTS

Residual solvents — Determine by gas chromatography in Volume 4 (under "Analytical Techniques, Chromatography").

Chromatography conditions
Column: 25% Diphenyl-75% dimethylpolysiloxane (60 m x 0.25 mm i.d., 0.25 μm film) [Aquatic-2 (GL-Sciences Inc.) or equivalent]
Carrier gas: Helium
Flow rate: 1.5 ml/min
Detector: Flame-ionization detector (FID)
Temperatures:
- injector: 280°
- column: Hold for 6 min at 40°, then 40-110° at 4°/min, 110-250° at 25°/min, hold for 10 min at 250°
- detector: 250°

Standard solutions
Solvent standard solution: Transfer 100 mg each of chromatography grade ethanol and isopropanol into a 100-ml volumetric flask containing about 90 ml water and dilute to 100 ml with water.
TBA standard solution: Transfer 100 mg of chromatography grade tertiary-butyl alcohol (TBA) into a 100-ml volumetric flask containing about 90 ml water and dilute to 100 ml with water.
Mixed standard solutions: Transfer 1, 2, 3, 4 and 5 ml of Solvent standard solution into each of five 100-ml volumetric flasks. Add 4 ml of TBA standard solution to each flask and dilute to volume with water.

Sample preparation
Disperse 1 ml of a suitable antifoam emulsion, such as Dow-Corning G-10 or equivalent, in 200 ml of water contained in a 1000-ml 24/40 round-bottom distilling flask. Add about 4 g of the sample, accurately weighed, and shake for 1 h on a wrist-action mechanical shaker. Connect the flask to a fractionating column, and distil about 95 ml, adjusting the heat so that foam does not enter the column. Add 4 ml of TBA standard solution to the distillate and make up to 100 ml with water to obtain the Sample solution.

Standard curves
Inject 1 μl of each Mixed standard solution into the chromatograph. Measure the peak areas for each solvent and TBA. Construct the standard curves by plotting the ratios of the peak areas of each of the solvents/TBA against the concentrations of each solvent (mg/ml) in the Mixed standard solutions.

Procedure
Inject 1 μl of the Sample solution into the chromatograph. Measure the peak areas for each solvent and TBA. Calculate the ratios of the peak areas of each solvent/TBA, and obtain the concentration of each solvent from the standard curves.

Calculate the percentage of each solvent from:

$$\% \text{ Solvent} = (C \times 100 / W \times 1000) \times 100$$

where  C is the concentration of solvent (mg/ml)
W is weight of sample (g)

# CAROB BEAN GUM (CLARIFIED)

*Prepared at the 69th JECFA (2008), published in FAO JECFA Monographs 5 (2008), superseding tentative specifications prepared at the 67th JECFA (2006) and published in FAO JECFA Monographs 3 (2006). An ADI "not specified" was established at the 25th JECFA (1981) for carob bean gum.*

**SYNONYMS** Locust bean gum clarified, INS No. 410

**DEFINITION** Primarily the ground endosperm of the seeds from *Ceratonia siliqua* (L.) Taub. (Fam. *Leguminosae*) mainly consisting of high molecular weight (approximately 50,000-3,000,000) polysaccharides composed of galactomannans; the mannose:galactose ratio is about 4:1. The seeds are dehusked by treating the kernels with dilute sulfuric acid or with thermal mechanical treatments, elimination of the germ, followed by milling and screening of the endosperm to obtain native carob bean gum. The gum is clarified by dispersing in hot water, filtration and precipitation with ethanol or isopropanol, filtering, drying and milling. The clarified carob bean gum does not contain cell wall materials. Clarified carob bean gum in the market is normally standardized with sugars for viscosity and reactivity.

C.A.S. number 9000-40-2

Structural formula

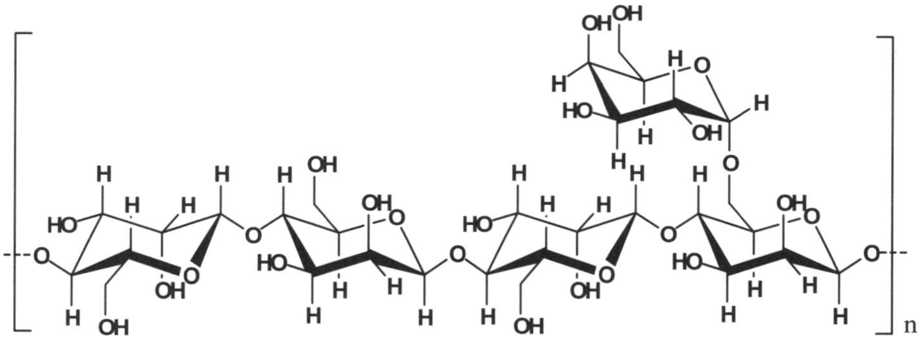

**DESCRIPTION** White to yellowish white, nearly odourless powder

**FUNCTIONAL USES** Stabilizer, thickener, emulsifier, gelling agent

**CHARACTERISTICS**

IDENTIFICATION

Solubility (Vol. 4) Insoluble in ethanol

Gel formation Add small amounts of sodium borate TS to an aqueous dispersion of the sample; a gel is formed.

Viscosity Transfer 2 g of the sample into a 400-ml beaker and moisten thoroughly with about 4 ml of isopropanol. Add 200 ml of water with vigorous stirring until the gum is completely and uniformly dissolved. An opalescent, slightly viscous solution is formed. Transfer 100 ml of this solution into another 400-ml beaker. Heat the mixture in a boiling water bath for about 10 min and cool to room temperature. There is an appreciable increase in viscosity (differentiating carob bean gums from guar gums).

| | |
|---|---|
| Gum constituents (Vol. 4) | Proceed as directed under Gum Constituents Identification using 100 mg of the sample instead of 200 mg and 1 to 10 µl of the hydrolysate instead of 1 to 5 µl. Use galactose and mannose as reference standards. These constituents should be present. |
| PURITY | Not more than 1 mg/kg<br>Determine using an AAS/ICP-AES technique appropriate to the specified level. The selection of sample size and method of sample preparation may be based on the principles of the methods described in Volume 4 (under "General Methods, Metallic Impurities"). |
| Loss on drying (Vol. 4) | Not more than 14% (105°, 5 h) |
| Total ash (Vol. 4) | Not more than 1.2% (800°, 3-4 h) (second peak). |
| Acid-insoluble matter (Vol. 4) | Not more than 3.5% |
| Protein (Vol. 4) | Not more than 1.0%<br>Proceed as directed under Nitrogen Determination (Kjeldahl Method) in Volume 4 (under "General Methods, Inorganic components"). The percentage of nitrogen determined multiplied by 6.25 gives the percentage of protein in the sample. |
| Starch | To a 1 in 10 solution of the sample add a few drops of iodine TS; no blue colour is produced |
| Residual solvents | Not more than 1% of ethanol or isopropanol, singly or in combination<br>See description under TESTS |
| Lead (Vol. 4) | Not more than 2 mg/kg<br>Determine using an AAS/ICP-AES technique appropriate to the specified level. The selection of sample size and method of sample preparation may be based on the principles of the methods described in Volume 4 (under "General Methods, Metallic Impurities"). |

Microbiological criteria (Vol. 4)

Initially prepare a $10^{-1}$ dilution by adding a 50 g sample to 450 ml of Butterfield's phosphate-buffered dilution water and homogenising the mixture in a high-speed blender.

Total (aerobic) plate count: Not more than 5,000 CFU/g
*E. coli:* Negative in 1 g
*Salmonella*: Negative in 25 g
Yeasts and moulds: Not more than 500 CFU/g

## TESTS
PURITY TESTS

Residual solvents

Determine by gas chromatography in Volume 4 (under "Analytical Techniques, Chromatography").

Chromatography conditions
Column: 25% Diphenyl-75% dimethylpolysiloxane (60 m x 0.25 mm i.d., 0.25 μm film) [Aquatic-2 (GL-Sciences Inc.) or equivalent]
Carrier gas: Helium
Flow rate: 1.5 ml/min
Detector: Flame-ionization detector (FID)
Temperatures:
- injector: 280°
- column: Hold for 6 min at 40°, then 40-110° at 4°/min, 110-250° at 25°/min, hold for 10 min at 250°
- detector: 250°

Standard solutions
Solvent standard solution: Transfer 100 mg each of chromatography grade ethanol and isopropanol into a 100-ml volumetric flask containing about 90 ml water and dilute to 100 ml with water.
TBA standard solution: Transfer 100 mg of chromatography grade tertiary-butyl alcohol (TBA) into a 100-ml volumetric flask containing about 90 ml water and dilute to 100 ml with water.
Mixed standard solutions: Transfer 1, 2, 3, 4 and 5 ml of Solvent standard solution into each of five 100-ml volumetric flasks. Add 4 ml of TBA standard solution to each flask and dilute to volume with water.

Sample preparation
Disperse 1 ml of a suitable antifoam emulsion, such as Dow-Corning G-10 or equivalent, in 200 ml of water contained in a 1000-ml 24/40 round-bottom distilling flask. Add about 4 g of the sample, accurately weighed, and shake for 1 h on a wrist-action mechanical shaker. Connect the flask to a fractionating column, and distil about 95 ml, adjusting the heat so that foam does not enter the column. Add 4 ml of TBA standard solution to the distillate and make up to 100 ml with water to obtain the Sample solution.

Standard curves
Inject 1 μl of each Mixed standard solution into the chromatograph. Measure the peak areas for each solvent and TBA. Construct the standard curves by plotting the ratios of the peak areas of each of the solvents/TBA against the concentrations of each solvent (mg/ml) in the Mixed standard solutions.

## Procedure

Inject 1 µl of the Sample solution into the chromatograph. Measure the peak areas for each solvent and TBA. Calculate the ratios of the peak areas of each solvent/TBA, and obtain the concentration of each solvent from the standard curves.

Calculate the percentage of each solvent from:

$$\% \text{ Solvent} = (C \times 100/W \times 1000) \times 100$$

where  C is the concentration of solvent (mg/ml)
W is weight of sample (g)

# ETHYL LAUROYL ARGINATE

*New specifications prepared at the 69th JECFA (2008), published in FAO JECFA Monographs 5 (2008). An ADI of 0-4 mg/kg bw was established at the 69th JECFA (2008).*

**SYNONYMS** Lauric arginate ethyl ester, lauramide arginine ethyl ester, ethyl-N$^\alpha$-lauroyl-L-arginate·HCl, LAE, INS No. 243

**DEFINITION** Ethyl lauroyl arginate is synthesised by esterifying arginine with ethanol, followed by reacting the ester with lauroyl chloride. The resultant ethyl lauroyl arginate is recovered as hydrochloride salt and is a white, solid product which is filtered off and dried.

Chemical name: Ethyl-N$^\alpha$-dodecanoyl-L-arginate·HCl

C.A.S. number: 60372-77-2

Chemical formula: $C_{20}H_{41}N_4O_3Cl$

Structural formula:

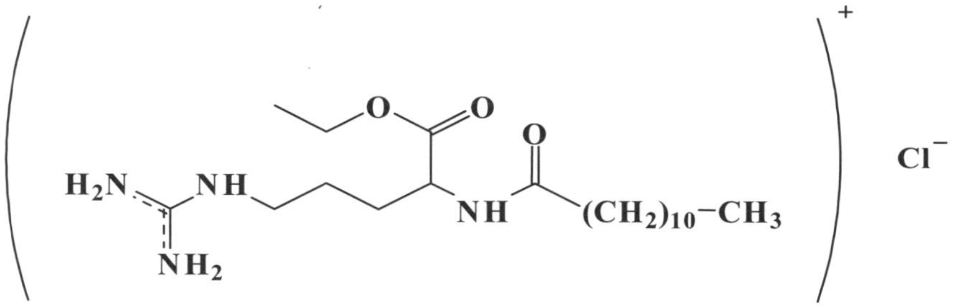

$$\left( H_2N-C(NH_2)-NH-(CH_2)_3-CH(COO-CH_2CH_3)-NH-CO-(CH_2)_{10}-CH_3 \right)^+ Cl^-$$

Formula weight: 421.02

Assay: Not less than 85% and not more than 95%

**DESCRIPTION** White powder

**FUNCTIONAL USES** Preservative

**CHARACTERISTICS**

IDENTIFICATION

pH (Vol.4): 3.0-5.0 (1% solution)

Solubility (Vol. 4): Freely soluble in water, ethanol, propylene glycol and glycerol

| | |
|---|---|
| Chromatography | The retention time for the major peak in a HPLC chromatogram of the sample is approx. 4.3 min using the conditions described in the Method of Assay. |
| **PURITY** | |
| Total ash (Vol. 4) | Not more than 2% (700°) |
| Water (Vol. 4) | Not more than 5% (Karl Fischer Titrimetric Method, "General Methods, Inorganic Components") |
| $N^\alpha$-Lauroyl-L-arginine | Not more than 3%<br>See description under TESTS |
| Lauric acid | Not more than 5%<br>See description under TESTS |
| Ethyl laurate | Not more than 3%<br>See description under TESTS |
| L-Arginine·HCl | Not more than 1%<br>See description under TESTS |
| Ethyl arginate·2HCl | Not more than 1%<br>See description under TESTS |
| Lead (Vol. 4) | Not more than 1 mg/kg<br>Determine using an AAS/ICP-AES technique appropriate to the specified level. The selection of sample size and method of sample preparation may be based on the principles of the methods described in Volume 4 (under "General Methods, Metallic Impurities"). |

## TESTS

PURITY TESTS

$N^\alpha$-Lauroyl-L-arginine — Determine by HPLC in Volume 4 (under "Analytical Techniques, Chromatography") using the conditions described in the Method of Assay.
NOTE: The retention time of $N^\alpha$-lauroyl-L-arginine is approx. 2.2 min.

Calculate the percentage of $N^\alpha$-lauroyl-L-arginine in the test sample as follows:

$$\% \; N^\alpha\text{-Lauroyl-L-arginine} = \frac{C \; (\mu g/ml) \times 50 \; (ml)}{W \; (mg) \times 1000} \times 100$$

where:
  C= $N^\alpha$-lauroyl-L-arginine concentration detected ($\mu$g/ml)
  W= weight of sample (mg)

Lauric acid and ethyl laurate — Determine by HPLC in Volume 4 (under "Analytical Techniques, Chromatography") using the following conditions.

### Chromatography
Liquid chromatograph equipped with a spectrophotometric detector.
Column: Symmetry C18, 150 x 3.9 mm, 5μm (Waters) or equivalent
Column temperature: room temperature
Mobile phase: acetonitrile/water (85:15) containing 0.1% trifluoroacetic acid
Flow rate: 1 ml/min
Wavelength: 212 nm
Injection volume: 10 μl

### Standard solution
Weigh accurately about 125 mg of lauric acid standard and 75 mg ethyl laurate standard into a 50-ml volumetric flask. Dissolve and dilute with the mobile phase to obtain a solution of about 2500 μg/ml of lauric acid and 1500μg/ml of ethyl laurate. Take 5, 10 and 15 ml of the solution and dilute to 50 ml with mobile phase for the standard curves.

### Sample solution
Weigh accurately about 500 mg of test sample into a 50-ml volumetric flask. Dissolve and dilute to 50 ml with mobile phase.

### Procedure
Inject the standard and sample solutions into the chromatograph and measure their concentration (C μg/ml) from their peak area and their standard curves.
NOTE: The retention time of lauric acid is approx. 3.65 min and that of ethyl laurate is approx. 11.2 min.

Calculate their percentage in the test sample as follows:

$$\% \text{ Lauric acid or ethyl laurate} = \frac{C (\mu g/ml) \times 50 (ml)}{W (mg) \times 1000} \times 100$$

where:
  C= lauric acid or ethyl laurate concentration detected (μg/ml)
  W= weight of sample (mg)

**L-Arginine·HCl and ethyl arginate·2HCl**

Determine by HPLC in Volume 4 (under "Analytical Techniques, Chromatography") using the following conditions:
NOTE: Use deionized water

### Chromatography
Liquid chromatograph equipped with a post-column derivatization and a spectrophotometric detector.
Column and packing: μ Bondapack C18, 300 x 3.9 mm, 10μm (Waters) or equivalent

Mobile phase: A-B-C-D (1:1:1:1.5)
A: 15 mmole/l sodium heptanesulphonate, B: 27 mmole/l phosphoric acid solution, C: 3 mmole/l sodium di-hydrogen phosphate solution, D: methanol
Flow rate: 0.8 ml/min
Flow rate of reagent solution: 0.8 ml/min

Column temperature: 65°
Wavelength: 340 nm
Injection volume: 10 µl

Standard solution
L-Arginine·HCl: Weigh accurately about 40 mg of L-arginine·HCl standard into a 100-ml volumetric flask. Dissolve and dilute to 100 ml with water to obtain a solution of about 400 µg/ml of L-arginine·HCl. Ethyl arginate·2HCl: Weigh accurately about 40 mg of ethyl arginate·2HCl standard into a 100-ml volumetric flask. Dissolve and dilute to 100 ml with water to obtain a solution of about 400 µg/ml of ethyl arginate·2HCl.
Take 2, 4, 6 and 8 ml of each solution and dilute to 25 ml with mobile phase separately for the standard curves.

Sample solution
Weigh accurately about 200 mg of test sample into a 25-ml volumetric flask. Dissolve and dilute to 25 ml with water.

Derivatizing solution
Mix 1 liter of 0.2M borate buffer solution (pH 9.4) with 0.8 g of *o*-phthaldialdehyde dissolved in 5 ml of methanol and 2 ml of 2-mercaptoethanol. The solution is stable 48 h at room temperature and without additional preventive measure but It is advisable to keep the solution under nitrogen and to prepare it freshly every 24-48 h.

Procedure
Inject the standard and sample solutions into the chromatograph and measure the area of the peak.
NOTE: The retention time of L-arginine·HCl is approx. 5.03 min and ethyl arginate·2HCl is approx. 6.70 min.

Calculate the percentage of L-arginine·HCl and ethyl arginate·2HCl in the test sample as follows:

$$\% \text{ L-Arginine·HCl or ethyl arginate·2HCl} = \frac{C\ (\mu g/ml) \times 50\ (ml)}{W\ (mg) \times 1000} \times 100$$

where:
C= L-arginine·HCl and ethyl arginate·2HCl concentration detected (µg/ml)
W= weight of sample (mg)

**METHOD OF ASSAY**  Determine by HPLC in Volume 4 (under "Analytical Techniques, Chromatography") using the following conditions:
NOTE: Use deionized water

Standards
Ethyl-$N^\alpha$-lauroyl-L-arginate·HCl standard
$N^\alpha$-lauroyl-L-arginine standard
(available from Laboratorios Miret, S.A, Géminis 4, Políg. Ind. Can Parellada, 08228 Terrassa, Spain)

### Chromatography
Liquid chromatograph equipped with a spectrophotometric detector.
Column and packing: Symmetry C18, 150 x 3.9 mm, 5μm (Waters) or equivalent
Column temperature: room temperature
Mobile phase: acetonitrile/water (50:50) containing 0.1% trifluoroacetic acid
Flow rate: 1 ml/min
Wavelength: 215 nm
Injection volume: 10 μl

### Standard solution
Weigh accurately about 25 mg of $N^{\alpha}$-lauroyl-L-arginine standard into a 25-ml volumetric flask. Dissolve and dilute to 25 ml with mobile phase (solution A). Weigh accurately about 150 mg of ethyl-$N^{\alpha}$-lauroyl-L-arginate·HCl standard into a 50-ml volumetric flask and dissolve with some milliliters of the mobile phase. Then, add 5 ml of solution A and dilute to 50 ml with mobile phase to obtain a solution of about 3000 μg/ml of ethyl-$N^{\alpha}$-lauroyl-L-arginate·HCl and 100 μg/ml of $N^{\alpha}$-lauroyl-L-arginine (solution B). Take 2, 4, 6, 8 and 10 ml of solution B and dilute to 25 ml with mobile phase for the standard curves.

### Sample solution
Weigh accurately about 50 mg of test sample into a 50-ml volumetric flask. Dissolve and dilute to 50 ml with mobile phase.

### Procedure
Inject the standard and sample solutions into the chromatograph and measure the area of the peak.
Note: The retention time of ethyl-$N^{\alpha}$-lauroyl-L-arginate·HCl is approx. 4.3 min.

Calculate the percentage of ethyl-$N^{\alpha}$-lauroyl-L-arginate·HCl in the test sample as follows:

$$\% \text{ Ethyl-}N^{\alpha}\text{-lauroyl-L-arginate·HCl} = \frac{C\ (\mu g/ml) \times 50\ (ml)}{W\ (mg) \times 1000} \times 100$$

where:
    C= ethyl-$N^{\alpha}$-lauroyl-L-arginate·HCl concentration detected (μg/ml)
    W= weight of sample (mg)

# GUAR GUM

*Prepared at the 69th JECFA (2008), published in FAO JECFA Monographs 5 (2008), superseding tentative specifications prepared at the 67th JECFA (2006) and published in FAO JECFA Monographs 3 (2006). An ADI "not specified" was established at the 19th JECFA (1975).*

| | |
|---|---|
| **SYNONYMS** | Gum cyamopsis, guar flour; INS No. 412 |
| **DEFINITION** | Primarily the ground endosperm of the seeds from Cyamopsis tetragonolobus (L.) Taub. (Fam. Leguminosae) mainly consisting of high molecular weight (50,000-8,000,000) polysaccharides composed of galactomannans; the mannose:galactose ratio is about 2:1. The seeds are crushed to eliminate the germ, the endosperm is dehusked, milled and screened to obtain the ground endosperm (native guar gum). The gum may be washed with ethanol or isopropanol to control the microbiological load (washed guar gum). |
| C.A.S. number | 9000-30-0 |

Structural formula

[chemical structure of guar gum polysaccharide repeating unit, shown with subscript n]

| | |
|---|---|
| **DESCRIPTION** | White to yellowish-white, nearly odourless, free-flowing powder |
| **FUNCTIONAL USES** | Thickener, stabilizer, emulsifier |

**CHARACTERISTICS**

IDENTIFICATION

| | |
|---|---|
| Solubility (Vol. 4) | Insoluble in ethanol |
| Gel formation | Add small amounts of sodium borate TS to an aqueous dispersion of the sample; a gel is formed. |
| Viscosity | Transfer 2 g of the sample into a 400-ml beaker and moisten thoroughly with about 4 ml of isopropanol. Add 200 ml of water with vigorous stirring until the gum is completely and uniformly dispersed. An opalescent, viscous solution is formed. Transfer 100 ml of this solution into another 400-ml beaker, heat the mixture in a boiling water bath for about 10 min and cool to room temperature. There is no substantial increase in viscosity (differentiating guar gums from carob bean gums). |

| | |
|---|---|
| Gum constituents (Vol. 4) | Proceed as directed under Gum Constituents Identification using 100 mg of the sample instead of 200 mg and 1 to 10 µl of the hydrolysate instead of 1 to 5 µl. Use galactose and mannose as reference standards. These constituents should be present. |
| Microscopic examination | Place some ground sample in an aqueous solution containing 0.5% iodine and 1% potassium iodide on a glass slide and examine under a microscope. Guar gum shows close groups of round to pear formed cells, their contents being yellow to brown. |

PURITY

| | |
|---|---|
| Loss on drying (Vol. 4) | Not more than 15.0% (105°, 5 h) |
| Borate | Absent by the following test<br>Disperse 1 g of the sample in 100 ml of water. The dispersion should remain fluid and not form a gel on standing. Mix 10 ml of dilute hydrochloric acid with the dispersion, and apply one drop of the resulting mixture to turmeric paper. No brownish red colour is formed. |
| Total ash (Vol. 4) | Not more than 1.5% (800°, 3-4 h) |
| Acid-insoluble matter (Vol. 4) | Not more than 7.0% |
| Protein (Vol. 4) | Not more than 10.0%<br>Proceed as directed under Nitrogen Determination (Kjeldahl Method) in Volume 4 (under "General Methods, Inorganic components"). The percentage of nitrogen determined multiplied by 6.25 gives the percentage of protein in the sample. |
| Residual solvents | Not more than 1% of ethanol or isopropanol, singly or in combination<br>See description under TESTS |
| Lead (Vol. 4) | Not more than 2 mg/kg<br>Determine using an AAS/ICP-AES technique appropriate to the specified level. The selection of sample size and method of sample preparation may be based on the principles of the methods described in Volume 4 (under "General Methods, Metallic Impurities"). |
| Microbiological criteria (Vol. 4) | Initially prepare a $10^{-1}$ dilution by adding a 50 g sample to 450 ml of Butterfield's phosphate-buffered dilution water and homogenizing the mixture in a high-speed blender.<br><br>Total (aerobic) plate count : Not more than 5,000 CFU/g<br>E. coli: Negative in 1g<br>Salmonella: Negative in 25g<br>Yeasts and moulds: Not more than 500 CFU/g |

# TESTS

PURITY TESTS

Residual solvents

Determine by gas chromatography in Volume 4 (under "Analytical Techniques, Chromatography").

Chromatography conditions
Column: 25% Diphenyl-75% dimethylpolysiloxane (60 m x 0.25 mm i.d., 0.25 μm film) [Aquatic-2 (GL-Sciences Inc.) or equivalent]
Carrier gas: Helium
Flow rate: 1.5 ml/min
Detector: Flame-ionization detector (FID)
Temperatures:
- injector: 280°
- column: Hold for 6 min at 40°, then 40-110° at 4°/min, 110-250° at 25°/min, hold for 10 min at 250°
- detector: 250°

Standard solutions
Solvent standard solution: Transfer 100 mg each of chromatography grade ethanol and isopropanol into a 100-ml volumetric flask containing about 90 ml water and dilute to 100 ml with water.
TBA standard solution: Transfer 100 mg of chromatography grade tertiary-butyl alcohol (TBA) into a 100-ml volumetric flask containing about 90 ml water and dilute to 100 ml with water.
Mixed standard solutions: Transfer 1, 2, 3, 4 and 5 ml of Solvent standard solution into each of five 100-ml volumetric flasks. Add 4 ml of TBA standard solution to each flask and dilute to volume with water.

Sample preparation
Disperse 1 ml of a suitable antifoam emulsion, such as Dow-Corning G-10 or equivalent, in 200 ml of water contained in a 1000-ml 24/40 round-bottom distilling flask. Add about 4 g of the sample, accurately weighed, and shake for 1 h on a wrist-action mechanical shaker. Connect the flask to a fractionating column, and distil about 95 ml, adjusting the heat so that foam does not enter the column. Add 4 ml of TBA standard solution to the distillate and make up to 100 ml with water to obtain the Sample solution.

Standard curves
Inject 1 μl of each Mixed standard solution into the chromatograph. Measure the peak areas for each solvent and TBA. Construct the standard curves by plotting the ratios of the peak areas of each of the solvents/TBA against the concentrations of each solvent (mg/ml) in the Mixed standard solutions.

Procedure
Inject 1 μl of the Sample solution into the chromatograph. Measure the peak areas for each solvent and TBA. Calculate the ratios of the peak areas of each solvent/TBA, and obtain the concentration of each solvent from the standard curves.

Calculate the percentage of each solvent from:

% Solvent = (C x 100/W x 1000) x 100

where  C is the concentration of solvent (mg/ml)
       W is weight of sample (g)

# GUAR GUM (CLARIFIED)

*Prepared at the 69th JECFA (2008), published in FAO JECFA Monographs 5 (2008), superseding tentative specifications prepared at the 67th JECFA (2006) and published in FAO JECFA Monographs 3 (2006). An ADI "not specified" was established at the 19th JECFA (1975) for guar gum.*

**SYNONYMS**  INS No. 412

**DEFINITION**  Primarily the ground endosperm of the seeds from Cyamopsis tetragonolobus (L.) Taub. (Fam. Leguminosae) mainly consisting of high molecular weight (50,000-8,000,000) polysaccharides composed of galactomannans; the mannose:galactose ratio is about 2:1. The seeds are crushed to eliminate the germ, the endosperm is dehusked, milled and screened to obtain the ground endosperm (native guar gum). The gum is clarified by dissolution in water, filtration and precipitation with ethanol or isopropanol. Clarified guar gum does not contain cell wall materials. Clarified guar gum in the market is normally standardized with sugars.

C.A.S. number  *9000-30-0*

Structural formula

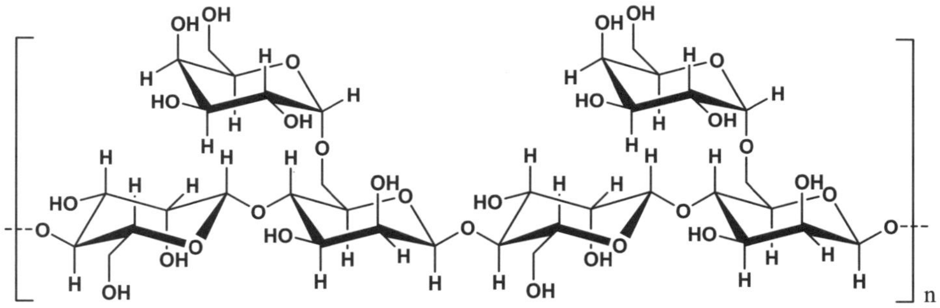

**DESCRIPTION**  White to yellowish white, nearly odourless, free-flowing powder

**FUNCTIONAL USES**  Thickener, stabilizer, emulsifier

**CHARACTERISTICS**

IDENTIFICATION

Solubility (Vol. 4)  Insoluble in ethanol

Gel formation  Add small amounts of sodium borate TS to an aqueous solution of the sample; a gel is formed.

Viscosity  Transfer 2 g of the sample into a 400-ml beaker and moisten thoroughly with about 4 ml of isopropanol. Add 200 ml of water with vigorous stirring until the gum is completely and uniformly dispersed. An opalescent, viscous solution is formed. Transfer 100 ml of this solution into another 400-ml beaker, heat the mixture in a boiling water bath for about 10 min and cool to room temperature. There is no substantial increase in viscosity

| | |
|---|---|
| Gum constituents (Vol. 4) | (differentiating guar gums from carob bean gums). Proceed as directed under Gum Constituents Identification using 100 mg of the sample instead of 200 mg and 1 to 10 µl of the hydrolysate instead of 1 to 5 µl. Use galactose and mannose as reference standards. These constituents should be present. |

PURITY

| | |
|---|---|
| Loss on drying (Vol. 4) | Not more than 15.0% (105°, 5 h) |
| Borate | Absent by the following test<br>Disperse 1 g of the sample in 100 ml of water. The dispersion should remain fluid and not form a gel on standing. Mix 10 ml of dilute hydrochloric acid with the dispersion, and apply one drop of the resulting mixture to turmeric paper. No brownish red colour is formed. |
| Total ash (Vol. 4) | Not more than 1.0% (800°, 3-4 h) |
| Acid-insoluble matter (Vol. 4) | Not more than 1.2% |
| Protein (Vol. 4) | Not more than 1.0%<br>Proceed as directed under Nitrogen Determination (Kjeldahl Method) in Volume 4 (under "General Methods, Inorganic components"). The percentage of nitrogen determined multiplied by 6.25 gives the percentage of protein in the sample. |
| Residual solvents | Not more than 1% of ethanol or isopropanol, singly or in combination<br>See description under TESTS |
| Lead (Vol. 4) | Not more than 2 mg/kg<br>Determine using an AAS/ICP-AES technique appropriate to the specified level. The selection of sample size and method of sample preparation may be based on the principles of the methods described in Volume 4 (under "General Methods, Metallic Impurities"). |
| Microbiological criteria (Vol. 4) | Initially prepare a $10^{-1}$ dilution by adding a 50 g sample to 450 ml of Butterfield's phosphate-buffered dilution water and homogenizing the mixture in a high-speed blender.<br><br>Total (aerobic) plate count: Not more than 5,000 CFU/g<br>E. coli: Negative in 1g<br>Salmonella: Negative in 25g<br>Yeasts and moulds: Not more than 500 CFU/g |

## TESTS

PURITY TESTS

| | |
|---|---|
| Residual solvents | Determine by gas chromatography in Volume 4 (under "Analytical Techniques, Chromatography").<br><br>Chromatography conditions |

Column: 25% Diphenyl-75% dimethylpolysiloxane (60 m x 0.25 mm i.d., 0.25 μm film) [Aquatic-2 (GL-Sciences Inc.) or equivalent]
Carrier gas: Helium
Flow rate: 1.5 ml/min
Detector: Flame-ionization detector (FID)
Temperatures:
- injector: 280°
- column: Hold for 6 min at 40°, then 40-110° at 4°/min, 110-250° at 25°/min, hold for 10 min at 250°
- detector: 250°

Standard solutions
Solvent standard solution: Transfer 100 mg each of chromatography grade ethanol and isopropanol into a 100-ml volumetric flask containing about 90 ml water and dilute to 100 ml with water.
TBA standard solution: Transfer 100 mg of chromatography grade tertiary-butyl alcohol (TBA) into a 100-ml volumetric flask containing about 90 ml water and dilute to 100 ml with water.
Mixed standard solutions: Transfer 1, 2, 3, 4 and 5 ml of Solvent standard solution into each of five 100-ml volumetric flasks. Add 4 ml of TBA standard solution to each flask and dilute to volume with water.

Sample preparation
Disperse 1 ml of a suitable antifoam emulsion, such as Dow-Corning G-10 or equivalent, in 200 ml of water contained in a 1000-ml 24/40 round-bottom distilling flask. Add about 4 g of the sample, accurately weighed, and shake for 1 h on a wrist-action mechanical shaker. Connect the flask to a fractionating column, and distil about 95 ml, adjusting the heat so that foam does not enter the column. Add 4 ml of TBA standard solution to the distillate and make up to 100 ml with water to obtain the Sample solution.

Standard curves
Inject 1 μl of each Mixed standard solution into the chromatograph. Measure the peak areas for each solvent and TBA. Construct the standard curves by plotting the ratios of the peak areas of each of the solvents/TBA against the concentrations of each solvent (mg/ml) in the Mixed standard solutions.

Procedure
Inject 1 μl of the Sample solution into the chromatograph. Measure the peak areas for each solvent and TBA. Calculate the ratios of the peak areas of each solvent/TBA, and obtain the concentration of each solvent from the standard curves.

Calculate the percentage of each solvent from:

$$\% \text{ Solvent} = (C \times 100/W \times 1000) \times 100$$

where  C is the concentration of solvent (mg/ml)
       W is weight of sample (g)

# IRON OXIDES

*Prepared at the 69th JECFA (2008), published in FAO JECFA Monographs 5 (2008), superseding the specifications prepared at the 63rd JECFA (2004), published in the Combined Compendium of Food Additive Specifications, FAO JECFA Monographs 1 (2005). An ADI of 0-0.5 mg/kg bw was established at the 53rd JECFA (1999).*

**SYNONYMS**  Iron Oxide yellow: CI Pigment Yellow 42 and 43; CI(1975) No. 77492; INS No. 172(iii)
Iron Oxide Red: CI Pigment Red 101 and 102; CI (1975) No. 77491; INS No. 172(ii)
Iron Oxide Black: CI Pigment Black 11; CI (1975) No. 77499; INS No. 172(i)

**DEFINITION**  Iron oxides are produced from ferrous sulfate by heat soaking, removal of water, decomposition, washing, filtration, drying and grinding. They are produced in either anhydrous or hydrated forms. Their range of hues includes yellows, reds, browns and blacks. The food-quality iron oxides are primarily distinguished from technical grades by their comparatively low levels of contamination by other metals; this is achieved by the selection and control of the source of the iron or by the extent of chemical purification during the manufacturing process.

Chemical names
- Iron Oxide Yellow: Hydrated ferric oxide, hydrated iron (III) oxide
- Iron Oxide Red: Iron sesquioxide, anhydrous ferric oxide, anhydrous iron (III) oxide
- Iron Oxide Black: Ferroso ferric oxide, iron (II,III) oxide

C.A.S. number
- Iron Oxide Yellow: 51274-00-1
- Iron Oxide Red: 1309-37-1
- Iron Oxide Black: 1317-61-9

Chemical formula
- Iron Oxide Yellow: $FeO(OH) \cdot xH_2O$
- Iron Oxide Red: $Fe_2O_3$
- Iron Oxide Black: $FeO \cdot Fe_2O_3$

Formula weight
- 88.85 $FeO(OH)$
- 159.70 $Fe_2O_3$
- 231.55 $FeO \cdot Fe_2O_3$

Assay  Not less than 60% of iron

**DESCRIPTION**  Yellow, red, brown or black powder.

**FUNCTIONAL USES**  Colour

**CHARACTERISTICS**

IDENTIFICATION

Solubility (Vol. 4) — Insoluble in water and organic solvents; soluble in concentrated mineral acids

PURITY

Loss on drying (Vol. 4) — Iron Oxide Red : Not more than 1.0% (105°, 4 h)

Water-soluble matter — Not more than 1.0%
See description under TESTS

Arsenic (Vol. 4) — Not more than 3 mg/kg
Determine by the atomic absorption hydride technique. The selection of sample size and method of sample preparation may be based on the principles of the methods described in Volume 4 (under "General Methods, Metallic Impurities").

Cadmium (Vol. 4) — Not more than 1 mg/kg
Determine using an atomic absorption/ICP technique appropriate to the specified level. The selection of sample size and method of sample preparation may be based on the principles of the methods described in Volume 4 (under "General Methods, Metallic Impurities").

Lead (Vol. 4) — Not more than 10 mg/kg
Determine using an atomic absorption/ICP technique appropriate to the specified level. The selection of sample size and method of sample preparation may be based on the principles of the methods described in Volume 4 (under "General Methods, Metallic Impurities").

Mercury (Vol. 4) — Not more than 1 mg/kg
Determine by the cold vapour atomic absorption technique.

## TESTS

PURITY TESTS

Water-soluble matter — Weigh accurately 5.0 g of iron oxide, transfer to a 250 ml beaker, add 200 ml of water and boil for 5 minutes; stir to avoid bumping. Cool the mixture, transfer the contents to a 250 ml volumetric flask, rinse the beaker with 25 ml of water, adding the rinsings to the flask; bring to volume with water and mix. Allow the mixture to stand for 10 minutes and filter the solution. Transfer 100 ml of filtrate into a clean dry tared beaker and carefully evaporate the solution to dryness on a boiling water bath. Dry the residue at 105 -110° for 2 hours, cool the beaker with residue in a desiccator, weigh the beaker, and calculate the amount of residue.

$$\text{Water-soluble matter (\%)} = 250 \times W_R/W_S$$

where $W_R$ is the weight of residue (g) and $W_S$ is the weight of sample taken (g).

**METHOD OF ASSAY** Weigh accurately about 0.2 g of the sample, add 10 ml of 5 N hydrochloric acid, and heat cautiously to boiling in a 200-ml conical flask until the sample has dissolved. Allow to cool, add 6 to 7 drops of 30% hydrogen peroxide solution and again heat cautiously to boiling until all the excess hydrogen peroxide has decomposed (about 2-3 min). Allow to cool, add 30 ml of water and about 2 g of potassium iodide and allow to stand for 5 min. Add 30 ml of water and titrate with 0.1 N sodium thiosulfate adding starch TS as the indicator towards the end of the titration. Each ml of 0.1N sodium thiosulfate is equivalent to 5.585 mg of Fe (III).

# ISOMALT

*Prepared at the 69th JECFA (2008), published in FAO JECFA Monographs 5 (2008), superseding specifications prepared at the 46th JECFA (1996), published in the Combined Compendium of Food Additive Specifications, FAO JECFA Monographs 1 (2005). An ADI 'not specified' was established at the 29th JECFA (1985).*

**SYNONYMS** Hydrogenated isomaltulose; INS No. 953

**DEFINITION** A mixture of hydrogenated mono- and disaccharides whose principal components are the disaccharides:

Chemical names
6-O-alpha-D-Glucopyranosyl-D-sorbitol (1,6-GPS) and
1-O-alpha-D-Glucopyranosyl-D-mannitol dihydrate (1,1-GPM)

C.A.S. number 64519-82-0

Chemical formula
6-O-alpha-D-Glucopyranosyl-D-sorbitol: $C_{12}H_{24}O_{11}$
1-O-alpha-D-Glucopyranosyl-D-mannitol dihydrate: $C_{12}H_{24}O_{11} \cdot 2H_2O$

Structural formula

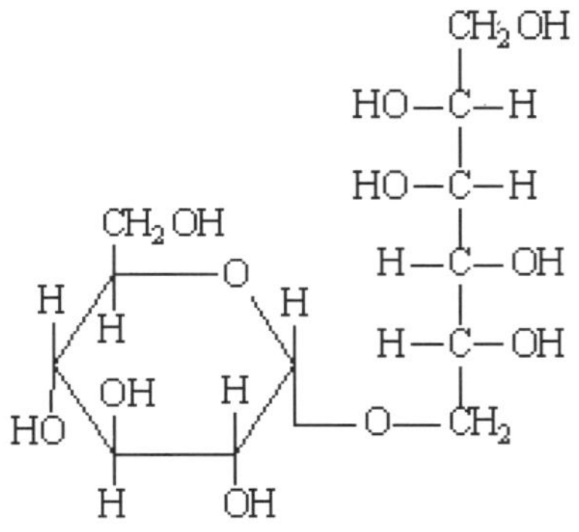

6-O-alpha-D-Glucopyranosyl-D-sorbitol

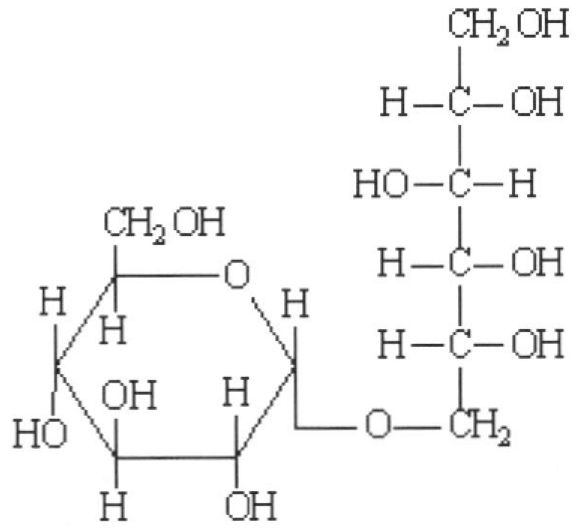

1-O-alpha-D-Glucopyranosyl-D-mannitol (without molecules of crystal water)

| | |
|---|---|
| Formula weight | 6-O-alpha-D-Glucopyranosyl-D-sorbitol: 344.32<br>1-O-alpha-D-Glucopyranosyl-D-mannitol dihydrate: 380.32 |
| Assay | Not less than 98% of hydrogenated mono- and disaccharides and not less than 86% of the mixture of 6-O-alpha-D-glucopyranosyl-D-sorbitol and 1-O-alpha-D-glucopyranosyl-D-mannitol on the anhydrous basis |

**DESCRIPTION**  Odourless, white, crystalline slightly hygroscopic substance

**FUNCTIONAL USES**  Sweetener, bulking agent, anticaking agent, glazing agent

**CHARACTERISTICS**

IDENTIFICATION

| | |
|---|---|
| Solubility (Vol. 4) | Soluble in water, very slightly soluble in ethanol |
| Thin layer chromatography (Vol. 4) | Passes test<br>See description under TESTS |

PURITY

| | |
|---|---|
| Water (Vol. 4) | Not more than 7.0% (Karl Fischer Titrimetric Method, "General Methods, Inorganic Components") |
| Sulfated ash (Vol. 4) | Not more than 0.05%<br>Test 5 g of the sample (Method I) |
| D-Mannitol | Not more than 3%<br>See Method of Assay |
| D-Sorbitol | Not more than 6%<br>See Method of Assay |
| Reducing sugars (Vol. 4) | Not more than 0.3%<br>Proceed as directed under *Reducing Substances (as glucose)*, Method II (under "General Methods, Organic Components"). The weight of cuprous oxide shall not exceed 50 mg. |
| Nickel (Vol. 4) | Not more than 2 mg/kg<br>Proceed as directed under *Nickel in Polyols* (under "General Methods, Inorganic Components"). |
| Lead (Vol. 4) | Not more than 1 mg/kg<br>Determine using an AAS/ICP-AES technique appropriate to the specified level. The selection of sample size and method of sample preparation may be based on the principles of the methods described in Volume 4 (under "General Methods, Metallic Impurities"). |

# TESTS

IDENTIFICATION TESTS

Thin layer chromatography

TLC plates
TLC aluminium foils or plates of approx. 12 cm length and coated with a layer of approx. 0.2 mm, Kieselgel 60 $F_{254}$, Art. 5554, Merck, or equivalent

Reference solution
Dissolve 500 mg of each of the following sugar alcohols in 100 ml of water: Sorbitol, mannitol, lactitol, maltitol, 1-O-alpha-D-gluco-pyranosyl-D-mannitol (1,1-GPM), and 6-O-alpha-D-glucopyranosyl-D-sorbitol (1,6-GPS)

Test solution
Dissolve 500 mg of sample in 100 ml of water

Solvent A
Isopropanol:n-butanol:aqueous boric acid solution (25 mg/ml):acetic acid:propionic acid (50:30:20:2:16;v/v)

Solvent B
Ethylacetate:pyridine:water:acetic acid:propionic acid (50:50:10:5:5;v/v)

Detecting solutions
I 0.1% Na-metaperiodate in water (w/w)
II ethanol:sulfuric acid:anisaldehyde:acetic acid (90:5:1:1;v/v)

Procedure
Apply approximately 0.3 µl each of the reference and test solution to the bottom of the TLC plate. Dry the spots in warm air. Develop the plate to a height of 10 cm in a developing chamber containing either solvent A or solvent B. Allow the plate to dry in warm air and dip the plate for up to 3 sec into Detecting solution I.

Dry the plate in hot air. Note: The plate should be completely dry on both sides. Dip the plate in Detecting solution II up to 3 sec and dry in hot air until coloured spots become visible. Optionally, the background colour may be brightened in warm steam.

The approximate $R_f$ values and colours of the spots on the TLC-plate of the substances specified above are described as "Compound / Colour / Solvent A($R_f$) / Solvent B($R_f$)". See below.

mannitol / reddish (light) / 0.36 / 0.40
sorbitol / brown / 0.36 / 0.36
GPM / blue-grey / 0.28 / 0.16
GPS / blue-grey / 0.25 / 0.13
maltitol / green / 0.26 / 0.22
lactitol / olive-green / 0.23 / 0.14

The $R_f$ values may vary slightly depending on the commercial source of the silica gel plates.

The principal spots in the chromatogram obtained from a test solution of isomalt are similar in $R_f$ value and colour to GPM and GPS.

PURITY TESTS

**METHOD OF ASSAY**

Internal standard solution
Dissolve suitable quantities of phenyl-ß-D-glucopyranoside and maltitol in water to obtain a solution of about 1 mg phenyl-ß-D-glucopyranoside and 50 mg maltitol per g water.

Standard solutions
Dissolve accurately weighed quantities of 1-O-alpha-D-glucopyranosyl-D-mannitol (1,1-GPM) and 6-O-alpha-D-glucopyranosyl-D-sorbitol (1,6-GPS), calculated as dry substance, in water to obtain two separate solutions having a concentration of about 50 mg per g each. Also prepare an aqueous standard solution containing approx. 1 mg mannitol and 1 mg sorbitol per g.

Sample solution
Dissolve an accurately weighed quantity of the sample (approx. 1 g) in water to obtain a concentration of about 10 g per 100 g.

Procedure
Pipet 100.0 mg of standard solution or sample solution into a glass tube fitted with a screw cap and add 100.0 mg of internal standard solution. Remove the water by lyophilization and dissolve the residue in 1.0 ml of pyridine. Add 4 mg O-benzyl-hydroxylamine hydrochloride, and cap the tube and set it aside for 12 h at room temperature. Then, add 1 ml of N-methyl-N-trimethylsilyl-trifluoroacetamide (MSTFA) and heat to 80° for 12 h shaking occasionally and allow to cool. Inject 1 µl portions of these solutions directly into a gas chromatograph under the following operating conditions:
- Column: Fused silica HT-8 (25 m x 0.22 mm x 0.25 µm), or equivalent
- Injector: Programmed temperature vaporizer: 30°; 270°/min to 300° (49 min)
- Detector: Flame ionization detector; 360°
- Temperature program: 80° (3 min); 10°/min to 210°; 5°/min to 350° (6 min)
- Carrier gas: Helium
- Flow rate: initial flow rate: approx. 1 ml/min at 80° and 1 atm; split flow: 25 ml/min

Approximate retention times
Hydrogenated monosaccharides:
Mannitol 19.5 min
Sorbitol 19.6 min
Internal standards:
Phenyl-ß-D-glucopyranoside 26.8 min
Maltitol 33.5 min
Hydrogenated disaccharides (32 - 36 min)
1,1-GPS 33.9 min
1,1-GPM 34.5 min

1,6-GPS 34.6 min

Calculate the percentages of the individual components, $w_I$, in the sample according to the following formula:

$$W_I (\%) = \frac{a_I \times m_S}{F_I \times a_S \times m_{ISOMALT}} \times 100$$

where
$a_I$ = peak area of component I (µV·s)
$a_S$ = peak area of internal standard (µV·s)
$m_S$ = mass of internal standard used for derivatization (mg d.s.)
$m_{ISOMALT}$ = mass of sample used for derivatization (mg d.s.)
$F_I$ = relative response factor $f_I/f_S$
$f_I$ = response factor of component I: $f_I = (a_I/m_I) \times (100/\% \text{ purity})$
$f_S$ = response factor of internal standard: $f_S = (a_S/m_S) \times (100/\% \text{ purity})$
$m_I$, $m_S$ = mass of component I or internal standard used for derivatization of standard sample (mg d.s.)

(NOTE: Use maltitol as internal standard for the calculation of hydrogenated disaccharides (e.g. 1,1-GPM, 1,6-GPS) and phenyl-ß-D-glucoside for the calculation of hydrogenated monosaccharides (mannitol, sorbitol). For the total of other saccharides (hydrogenated or not), subtract the sum of 1,1-GPM, 1,6-GPS, sorbitol and mannitol from 100%.)

# MONOMAGNESIUM PHOSPHATE

*Prepared at the 69th JECFA (2008), published in FAO JECFA Monographs 5 (2008), based on the previously withdrawn tentative specifications prepared at the 61st JECFA and published in FNP 52, Add 11 (2003). A group MTDI of 70 mg/kg bw, expressed as phosphorus from all food sources, was established at the 26th JECFA (1982).*

| | |
|---|---|
| **SYNONYMS** | Monomagnesium orthophosphate, Magnesium dihydrogen phosphate; Magnesium phosphate, monobasic; Magnesium biphosphate; Acid magnesium phosphate; INS No. 343(i) |
| **DEFINITION** | Monomagnesium phosphate is manufactured by partial neutralization of phosphoric acid with magnesium oxide and drying of the resultant product. |
| Chemical names | Monomagnesium dihydrogen phosphate |
| C.A.S. number | 13092-66-5 (Anhydrous)<br>15609-87-7 (Dihydrate) |
| Chemical formula | $Mg(H_2PO_4)_2 \cdot x H_2O$ (x = 0 to 4) |
| Formula weight | 218.3 (Anhydrous)<br>254.3 (Dihydrate)<br>290.3 (Tetrahydrate) |
| Assay | Not less than 96% and not more than 102% as $Mg_2P_2O_7$ on the ignited basis |
| **DESCRIPTION** | White, odourless, crystalline powder |
| **FUNCTIONAL USES** | Acidity regulator, nutrient |
| **CHARACTERISTICS** | |
| IDENTIFICATION | |
| Solubility (Vol. 4) | Slightly soluble in water |
| Magnesium (Vol. 4) | Passes test |
| Phosphate (Vol. 4) | Passes test |
| PURITY | |
| Loss on drying (Vol. 4) | Anhydrous: Not more than 1.5 % (105°, 4 h) |
| Lost of ignition (Vol. 4) | Anhydrous: Not more than 18.5 %<br>Dihydrate: Not more than 33 %<br>Tetrahydrate: Not more than 43% |
| | Accurately weigh about 2 g of sample, and ignite, preferably in a muffle furnace at about 800° for 30 min. Allow the crucible to cool |

in a desiccator to constant weight. Save the residue for the Assay.

| | |
|---|---|
| Fluoride (Vol. 4) | Not more than 10 mg/kg<br>See description under TESTS |
| Arsenic (Vol. 4) | Not more than 3 mg/kg<br>Determine by the atomic absorption hydride technique. The selection of sample size and method of sample preparation may be based on the principles of the methods described in Volume 4 (under "General Methods, Metallic Impurities"). |
| Lead (Vol. 4) | Not more than 4 mg/kg<br>Determine using an atomic absorption/ICP technique appropriate to the specified level. The selection of sample size and method of sample preparation may be based on the principles of the methods described in Volume 4 (under "General Methods, Metallic Impurities"). |

## TESTS

PURITY TESTS

| | |
|---|---|
| Fluoride (Vol. 4) | Use Method III. The standard curve constructed in Method III may not be suitable for samples containing low fluoride levels. Therefore, it will be necessary to prepare standard solutions with concentrations other than those specified for Method III for the construction of a standard curve and to choose a sample size that will bring the fluoride concentration within the standard curve. |
| **METHOD OF ASSAY** | Accurately weigh 200 mg of the residue obtained in the test for Loss on ignition in a high 250 ml beaker. Dissolve the residue in 2 ml of hydrochloric acid (16 %) and add 100 ml of water. Heat the solution to 50° to 60° and add 10 ml of 0.1 M disodium EDTA from a buret. Add a magnetic stirring bar and, while stirring, adjust with 1 N sodium hydroxide to pH 10. Add 10 ml of ammonia-ammonium chloride buffer TS (Vol. 4), 12 drops of Eriochrome black TS and continue the titration with 0.1 M disodium EDTA until the red colour changes to green. [NOTE: The solution must be clear when the end point is reached] Calculate the weight (mg) of $Mg_2P_2O_7$ in the residue taken by the formula |

$$9.14 \times V$$

where V is the volume (ml) of 0.1 M disodium EDTA required in the titration.

# PAPRIKA EXTRACT
## (TENTATIVE)

*New tentative specifications prepared at the 69$^{th}$ JECFA (2008), published in FAO JECFA Monographs 5 (2008). No ADI was allocated at the 69$^{th}$ JECFA (2008).*

*Information required on batches of commercially available products:*
- *analytical data on composition*
- *levels of capsaicinoids*
- *levels of arsenic*

**SYNONYMS**  INS No. 160c, Capsanthin, Capsorubin

**DEFINITION**  Paprika extract is obtained by solvent extraction of the dried ground fruit pods of *Capsicum annuum*. The major colouring compounds are capsanthin and capsorubin. Other coloured compounds, such as other carotenoids are also present. The balance of the extracted material is lipidic in nature and varies depending on the primary extraction solvent. Commercial preparations may be diluted and standardised with respect to colour content using refined vegetable oil.

Only methanol, ethanol, 2-propanol, acetone, hexane, ethyl acetate and supercritical carbon dioxide may be used as solvents in the extraction.

Chemical names

Capsanthin: (3R, 3'S, 5'R)-3,3'-dihydroxy-β,κ-carotene-6-one
Capsorubin: (3S, 3'S, 5R, 5'R)-3,3'-dihydroxy-κ,κ-carotene-6,6'-dione

C.A.S number

Capsanthin: 465-42-9
Capsorubin: 470-38-2

Chemical formula

Capsanthin: $C_{40}H_{56}O_3$
Capsorubin: $C_{40}H_{56}O_4$

Structural formula

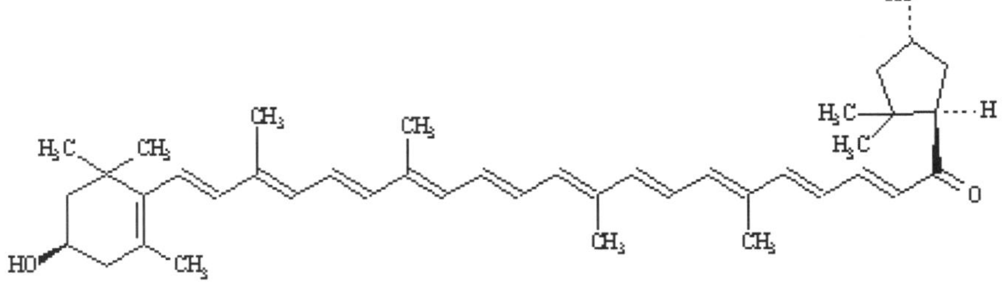

Capsanthin

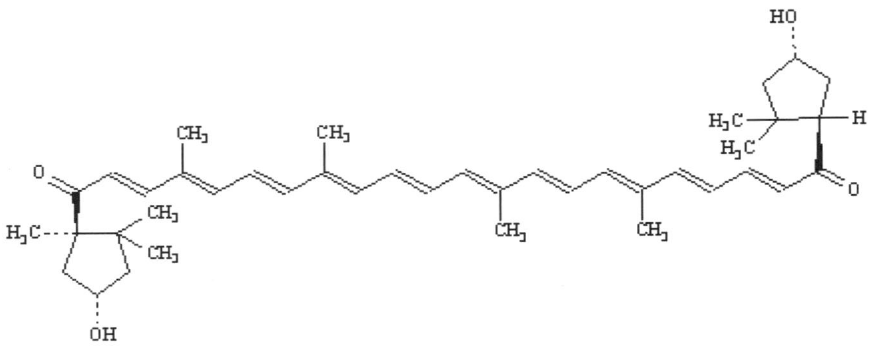

Capsorubin

| | |
|---|---|
| Formula weight | Capsanthin: 584.85<br>Capsorubin: 600.85 |
| Assay | Total carotenoids: not less than declared.<br>Capsanthin/capsorubin: Not less than 30% of total carotenoids. |

**DESCRIPTION**  Dark-red viscous liquid

**FUNCTIONAL USE**  Colour

**CHARACTERISTICS**

IDENTIFICATION

| | |
|---|---|
| Solubility | Practically insoluble in water, soluble in acetone |
| Spectrophotometry | Maximum absorption in acetone at about 462 nm and in hexane at about 470 nm. |
| Colour reaction | To one drop of sample add 2-3 drops of chloroform and one drop of sulfuric acid. A deep blue colour is produced. |
| High Performance Liquid Chromatography (HPLC) | Passes test.<br>See Method of assay, Capsanthin/capsorubin |

PURITY

| | |
|---|---|
| Residual solvents (Vol. 4) | Ethyl acetate, methanol, ethanol, acetone, 2-propanol, hexane: Not more than 50 mg/kg either singly or in combination |
| Capsaicinoids | *Information required on levels in commercial products*<br>See description under TESTS |
| Arsenic (Vol. 4) | Not more than 3 mg/kg<br>Determine by the atomic absorption hydride technique. The selection of sample size and method of sample preparation may be based on the principles of the methods described in Volume 4 (under "General Methods, Metallic Impurities"). |

Lead (Vol. 4)  Not more than 2 mg/kg
Determine using an atomic absorption/ICP technique appropriate to the specified level. The selection of sample size and method of sample preparation may be based on the principles of the methods described in Volume 4 (under "General Methods, Metallic Impurities").

## TESTS

PURITY TESTS

Capsaicinoids

Capsaicinoids are determined by reversed-phase HPLC (Volume 4 under "Chromatography") using a reference standard to allow quantification.

### Preparation of standard
Prepare all standard solutions in ethanol and keep out of direct sunlight.
- Standard solution A, 150 µg/ml: Accurately weigh and transfer 75 mg of N-vanillyl-n-nonenamide, >99 % (CAS Registry Number 2444-46-4) into a 500 ml volumetric flask, dissolve and dilute to volume. Mix thoroughly.
- Standard solution B, 15 µg/ml: Pipet 10 ml standard solution A into a 100 ml volumetric flask, dilute to volume, and mix well.
- Standard solution C, 0.75 µg/ml: Pipet 5 ml of standard solution B into 100 ml volumetric flask, dilute to volume, and mix well.

### Preparation of sample
Accurately weigh up to 5 g extract into a 50 ml volumetric flask, do not allow the extract to coat the sides of the flask. Add 5 ml acetone (ACS Grade) to the flask and swirl until the sample is completely dispersed. Ensure the extract has not coated the bottom of flask when neck is at a 45° angle. Slowly add ethanol (95% or denatured) with mixing until the solution becomes cloudy. Dilute to volume and mix well. Directly pipet 5 ml sample mixture into a 10 ml syringe attached to a 6 ml C-18 SEP-PAK cartridge. Take care to avoid coating of sample on the sides of syringe. Allow the aliquot to pass through the SEP-PAK and collect the eluent in a 25 ml volumetric flask. Rinse the SEP-PAK with three 5 ml portions of ethanol, and collect in the flask. Dilute to volume with ethanol and mix. Filter through a 0.45 µm syringe filter and collect in a glass vial.

### Apparatus
Liquid chromatograph equipped with a 20 µl sample loop injector, a fluorescence detector and/or ultraviolet detector and integrator.
Column: LC-18 (150 x 4.6 mm id, 5 µm)
Detector:
   Fluorescence - Excitation 280 nm and emission 325 nm
   UV Detector - 280 nm
Mobile phase: 40% acetonitrile and 60% deionised $H_2O$ containing 1% Acetic acid (v/v).
Flow rate: 1.5 ml/min

## Procedure

Inject 20 µl of the sample solution in duplicate. Inject the appropriate standard solution (Standard solution C is appropriate for samples expected to contain low levels of capsaicins) prior to the first sample injection and after every 6 sample injections. Purge the column with 100% acetonitrile for 30 min at 1.5 ml/min after no more than 30 sample injections. Equilibrate with mobile phase prior to further determinations.

## Calculations

Calculate individual capsaicinoids (µg/ml) as follows:

Nordihydrocapsaicin: $C_N = (N/A) \times (C_s/R_N)$
Capsaicin: $C_C = (C/A) \times (C_s/R_C)$
Dihydrocapsaicin: $C_D = (D/A) \times (C_s/R_D)$

Total capsaicins (µg/ml) = nordihydrocapsaicin + capsaicin + dihydrocapsaicin

where
- A = average peak area of standard;
- N, C, and D = average peak areas for respective capsaicinoids (nordihydrocapsaicin, capsaicin and dihydrocapsaicin) from duplicate injections;
- $C_s$ = concentration of std in µg/ml;
- $C_{N,C,D}$ = concentration of compound in extract expressed as µg/ml;
- $R_N$, $R_C$, and $R_D$ = response factors of respective capsaicinoids relative to standard.

Response factors:
Nordihydrocapsaicin (N) UV: $R_N = 0.98$; FLU: $R_N = 0.92$
Capsaicin (C) UV: $R_C = 0.89$; FLU: $R_C = 0.88$
Dihydrocapsaicin (D) UV: $R_D = 0.93$; FLU: $R_D = 0.93$
N-vanillyl-n-nonenamide UV: R = 1.00; FLU: R = 1.00
Relative retention times: Nordihydrocapsaicin 0.90; N-vanillyl-n-nonenamide 1.00, Capsaicin 1.00; Dihydrocapsaicin 1.58

## Capsanthin/capsorubin

Determine the total carotenoids in paprika extract by spectrophotometry.

Accurately weigh 300 to 500 mg of sample, and transfer quantitatively to a 100 ml volumetric flask. Dilute with acetone to volume, dissolve by shaking and leave to stand for 2 min. Pipet 1 ml of this extract into another 100 ml volumetric flask, dilute to volume with acetone, and shake well. Transfer a portion to the spectrophotometer cell, and read the absorbance A at 462 nm. Adjust the sample concentration to obtain an absorbance between 0.3 and 0.7.

Determine total pigment (%) as capsanthin and capsorubin

$$\text{Total} = \frac{a}{2100} \times \frac{10000}{W}$$

where
> $a$ = absorbance of sample
> $2100 = A^{1\%}_{1cm}$ for capsanthin/capsorubin in acetone at 462 nm
> $W$ = weight of sample (g)

Determine the identity and relative purity of paprika extract by reversed-phase HPLC. See Volume 4 under "Chromatography". The sample is saponified to release the parent hydroxy-carotenoids from the extracts prior HPLC analysis.

Sample preparation
Dissolve 0.2 g of the sample in acetone, quantitatively transfer into a 500 ml separatory funnel and add enough acetone to make up to 100 ml. Add 100 ml diethyl ether and mix well. Remove any insoluble particles by filtration. Add 100 ml of KOH-methanol (20%) and leave the solution for one hour. Shake periodically. Remove the aqueous phase and wash the organic phase several times with distilled water until the washings are neutral. Filter through a bed of anhydrous $Na_2SO_4$ and evaporate to dryness in a rotary evaporator at a temperature below 35°. Dissolve the pigments in acetone and make up to 25 ml in a volumetric flask. Keep the samples refrigerated until analysis by HPLC. Thoroughly disperse the samples, e.g. by sonication, and filter through a 0.45 µm filter before analysis.

Chromatography
Filter acetone (HPLC grade) and deionised water and de-gas before use.
Column: Reversed-phase C-18 (250 x 4 mm i.d.)
Precolumn: Reversed-phase C-18 (50 x 4 mm i.d.)
Mobile phase: Program a gradient acetone/water as follows:

| Time (min) | Acetone (%) | Water (%) |
|---|---|---|
| -10 (pre-injection) | 75 | 25 |
| 0 | 75 | 25 |
| 5 | 75 | 25 |
| 10 | 95 | 5 |
| 17 | 95 | 5 |
| 22 | 100 | 0 |
| 27 | 75 | 25 |

Flow rate: 1.5 ml/min
Detector: Diode array detector, store spectra in the range of 350-600 nm.
Detection wavelength: 450 nm
Injection volume: 5 µl

Identify peaks by comparing the peaks obtained with known standards and quantify the individual carotenoids. Saponified carotenoids will elute in the same order, with capsorubin and some minor carotenoids eluting first and β-carotene in last place. The order of elution is:

- Neoxanthin
- Capsorubin
- Violaxanthin
- Capsanthin
- Antheraxanthin
- Mutatoxanthin
- Cucurbitaxanthin A (Capsolutein)
- Zeaxanthin
- Cryptocapsin
- β-Cryptoxanthin
- β-Carotene

Calculate the percent of each peak using the total area of the peaks in the chromatogram. Sum the percentages of capsanthin and capsorubin to get the total value.

# PATENT BLUE V

*Prepared at the 69th JECFA (2008), published in FAO JECFA Monographs 5 (2008), superseding specifications prepared at the 31st JECFA (1987), published in the combined Compendium of Food Additive Specifications, FAO JECFA Monographs 1 (2005). No ADI could be allocated at the 26th JECFA (1982).*

**SYNONYMS**  CI Food Blue 5, Patent Blue 5; CI (1975) No. 42051; INS No. 131

**DEFINITION**  Patent Blue V consists essentially of the calcium or sodium salt of 2-[(4-diethylaminophenyl)(4-diethylimino-2,5-cyclohexadien-1-ylidene)methyl]-4-hydroxy-1,5-benzenedisulfonate and subsidiary colouring matters. Water, sodium chloride, sodium sulfate, calcium chloride, and calcium sulfate can be present as the principal uncoloured components.

Patent Blue V may be converted to the corresponding aluminium lake, in which case only the *General Specifications for Aluminium Lakes of Colouring Matters* applies.

Chemical names  Calcium or sodium salt of 2-[(4-diethylaminophenyl)(4-diethylimino-2,5-cyclohexadien-1-ylidene)methyl]-4-hydroxy-1,5-benzene-disulfonate; Calcium or sodium salt of [4-[*alpha*-(4-diethyl-aminophenyl)-5-hydroxy-2,4-disulfonatophenylmethylidene]-2,5-cyclohexadien-1-ylidene] diethylammonium hydroxide inner salt

C.A.S. number  3536-49-0

Chemical formula  Calcium salt: $C_{27}H_{31}N_2O_7S_2 \cdot \frac{1}{2}Ca$
Sodium salt: $C_{27}H_{31}N_2O_7S_2Na$

Structural formula

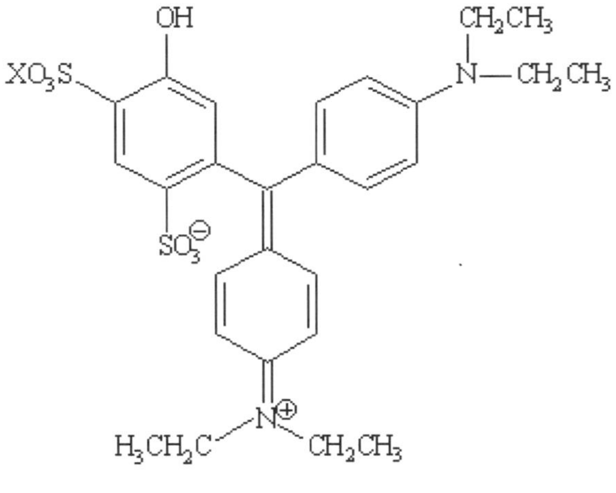

where
X = ½Ca for the calcium salt
X = Na for the sodium salt

Formula weight  ½Calcium salt: 579.14
Sodium salt: 582.15

Assay  Not less than 85% total colouring matter

| | |
|---|---|
| **DESCRIPTION** | Blue powder or granules |
| **FUNCTIONAL USES** | Colour |

**CHARACTERISTICS**

IDENTIFICATION

| | |
|---|---|
| Solubility (Vol. 4) | Soluble in water; slightly soluble in ethanol |
| Colouring matters, Identification (Vol. 4) | Passes test |

PURITY

| | |
|---|---|
| Water content (Loss on drying) (Vol. 4) | Not more than 15% together with chloride and sulfate calculated as sodium salts |
| Water-insoluble matter (Vol. 4) | Not more than 0.5% |
| Lead (Vol. 4) | Not more than 2 mg/kg<br>Determine using an AAS/ICP-AES technique appropriate to the specified level. The selection of sample size and method of sample preparation may be based on the principles of the methods described in Volume 4 (under "General Methods, Metallic Impurities"). |
| Chromium (Vol. 4) | Not more than 50 mg/kg<br>Determine using an AAS/ICP-AES technique appropriate to the specified level. The selection of sample size and method of sample preparation may be based on the principles of the methods described in Volume 4 (under "General Methods, Metallic Impurities") |
| Subsidiary colouring matter content (Vol. 4) | Not more than 2%<br>Use the following conditions:<br>Chromatography solvent: n-butanol:water:ethanol:ammonia (s.g. 0.880) (600:264:135:6)<br>Height of ascent of solvent front: approximately 17 cm |
| Organic compounds other than colouring matters | Not more than 0.5% (Sum of 3-hydroxybenzaldehyde, 3-hydroxybenzoic acid, 3-hydroxy-4-sulfonatobenzoic acid and *N,N*-diethylaminobenzenesulfonic acids)<br>See description under TESTS |
| Leuco base (Vol. 4) | Not more than 4%<br>Proceed as directed in Volume 4 using the following parameters:<br>- Sample: 110 mg<br>- Ratio of the formula weight of the colouring matter to the formula weight of its leuco base:<br>Sodium salt: 582.15/606.66 = 0.95960<br>½Calcium salt: 579.14/600.76 = 0.96401<br>- Absorptivity: 0.200 l/(mg·cm) at 638 nm |
| Unsulfonated primary aromatic amines (Vol. 4) | Not more than 0.01%, calculated as aniline |

Ether-extractable matter (Vol. 4) — Not more than 0.2%

## TESTS

PURITY TESTS

Organic compounds other than colouring matters (Vol. 4) — Proceed as directed under *Determination by High Performance Liquid Chromatography* using the following conditions:
Instrument: High Performance Liquid Chromatograph fitted with a gradient elution accessory

Detector: A UV detector monitored at 254 nm
Column: 250 x 4 mm (Kartusche). LiChrosorb RP 18, 7 μm or equivalent.
Mobile phase:
(A) Acetate buffer pH 4.6: water (10% w/v) - prepared using 1 M sodium hydroxide, 1 M acetic acid and water (5:10:35)
(B) Acetonitrile

Gradient

| Min | % (A) | % (B) | Flow rate (ml/min) |
|-----|-------|-------|--------------------|
| 0   | 85    | 15    | 1                  |
| 12  | 85    | 15    | 1                  |
| 25  | 20    | 80    | 2                  |
| 28  | 20    | 80    | 2                  |
| 40  | 85    | 15    | 1                  |

**METHOD OF ASSAY** — Proceed as directed under *Colouring Matters Content by Titration with Titanous Chloride* (Volume 4), under *Food Colours, Colouring Matters*), using the following:
Weight of sample: 1.3-1.4 g
Buffer: 15 g sodium hydrogen tartrate
Weight (*D*) of colouring matters equivalent to 1.00 ml of 0.1 N $TiCl_3$:
28.98 mg of the calcium salt
29.13 mg of the sodium salt.

## PHOSPHOLIPASE C EXPRESSED IN *PICHIA PASTORIS*

*New specifications prepared at the 69th JECFA (2008), published in FAO JECFA Monographs 5 (2008). An ADI "not specified" was established at the 69th JECFA (2008).*

| | |
|---|---|
| **SYNONYMS** | Phospholipase C; lecithinase C; lipophosphodiesterase C; phosphatidase C |
| **SOURCES** | Phospholipase C is produced by submerged fed-batch fermentation of a genetically modified strain of Pichia pastoris which contains the phospholipase C gene derived from a soil sample. The enzyme is recovered from the fermentation broth. The recovery process includes the separation of cellular biomass, clarification, ultrafiltration, diafiltration, and polish filtration. The final product is formulated using food-grade stabilizing and preserving agents and is standardized to the desired activity. |
| Active principles | Phospholipase C |
| Systematic names and numbers | Phosphatidylcholine cholinephosphohydrolase; EC 3.1.4.3; CAS No. 9001-86-9 |
| Reactions catalysed | Hydrolysis of phosphodiester bonds at the sn-3 position in glycerophospholipids including phosphatidylcholine, phosphatidylethanolamine, and phosphatidylserine to yield 1,2-diacylglycerol and the corresponding phosphate esters |
| Secondary enzyme activities | No significant levels of secondary enzyme activities. |
| **DESCRIPTION** | Yellow to brown liquid |
| **FUNCTIONAL USES** | Enzyme preparation.<br>Used in refining vegetable oils intended for human consumption. |
| **GENERAL SPECIFICATIONS** | Must conform to the latest edition of the JECFA General Specifications and Considerations for Enzyme Preparations Used in Food Processing. |

**CHARACTERISTICS**

IDENTIFICATION

| | |
|---|---|
| Phospholipase C activity | The sample shows phospholipase C activity.<br>See description under TESTS. |

**TESTS**

Enzyme activity

**Principle**
Phospholipase C catalyses the hydrolysis of phosphatidylcholine to 1,2-diacylglycerol and phosphorylcholine. Phosphorylcholine is subsequently titrated with potassium hydroxide. The activity of phospholipase C is determined by measuring the rate of

consumption of potassium hydroxide required to maintain pH 7.3 at 37°.

The enzyme activity is expressed in phospholipase C units (PLCU). One phospholipase C unit is defined as the quantity of the enzyme that will hydrolyse 1 µmol phosphatidylcholine per minute under standard conditions (pH=7.3; 37°).

**Apparatus**
Auto-titrator (Brinkmann Instruments, Titrandos® 835 or equivalent)
pH meter (Beckman Coulter, model F350 or equivalent)
Homogenizer (M133/1281-0, 2-speed, BioSpec Products, catalog # 1281, or equivalent)
Circulating water bath

**Reagents and solutions**
(NOTE: use deionized water)

*Potassium hydroxide (0.01 N)*: 0.01 N KOH certified titration reagent (Brinkmann Instruments 019091104 or equivalent). Use for titration of phosphorylcholine in the phospholipase C activity assay.

*Zinc sulfate solution (100 mM)*: Weigh 2.88 g of zinc sulfate heptahydrate (crystalline, certified ACS) and dissolve in water in a 100-ml volumetric flask. Add water to volume. The solution is stable for up to 30 days at room temperature.

*Calcium chloride solution (100 mM)*: Weigh 1.47 g of calcium chloride dihydrate (certified ACS) and dissolve in water in a 100-ml volumetric flask. Add water to volume. The solution is stable for up to 30 days at room temperature.

*Triton X-100 solution (approximately 10%)*: Weigh 10 g of Triton X-100 (Sigma-Aldrich T9284 or equivalent) into a 200-ml beaker. Add 100 ml of water and mix for at least 1 hr on a rotating table. The solution is stable in a closed container for up to 30 days at room temperature.

*Substrate solution (20 mM phosphatidylcholine, approximately 2.5% Triton X-100, 5 mM calcium chloride)*: Weigh 3.24 g of phosphatidylcholine (Phospholipon 90G (containing at least 94% phosphatidylcholine), American Lecithin Company or equivalent) into a 500 ml beaker. Add 50 ml of 10% Triton X-100 solution and 10.0 ml of 100 mM calcium chloride solution. Adjust volume to 200 ml with water and mix. Homogenize the solution using a hand-held homogenizer at low setting (7,000 rpm) for approx. 45 sec or until a uniform dispersion is obtained. Check the pH and, if necessary, adjust to the range of 6.5-7.0 using 0.2 N sodium hydroxide solution certified, Fisher Scientific SS274-1 or equivalent). The solution should be prepared on the day of testing.

*Dilution buffer (0.1% Triton X-100, 1 mM zinc sulfate, 1% gum arabic)*: Weigh 0.5 g of Triton X-100 and 5.0 g of gum arabic (Sigma-Aldrich G9752 or equivalent). Dissolve with stirring in 450 ml of water in a 1000 ml beaker. Add 5 ml of 100 mM zinc sulfate solution and adjust the pH to the range 7.0-7.2 using 0.2 N sodium hydroxide solution. Transfer to a 500 ml volumetric flask and add water to volume. The solution is stable for up to 30 days at 4°.

*Sample solution*: Weigh to ±0.1 mg approximately 1 g of the phospholipase C enzyme preparation into a 50 ml volumetric flask. Add the dilution buffer to volume and mix. Dilute with the dilution buffer to obtain a solution with an activity of approximately 12 PLCU/ml. The solution should be prepared on the day of testing.

**Procedure**
1. Program the titrator to maintain the pH 7.3 and measure the consumption of 0.01 N KOH in milliliters per minute.
2. Set the temperature of the recirculating water bath at 37°.
3. Calibrate the pH electrode at pH 4, 7, and 10.
4. Transfer 20 ml of the substrate solution into the water-jacketed titration vessel of 50 ml capacity connected to the recirculating water bath, cover with the lid and stir.
5. Allow the substrate solution to equilibrate to 37°.
6. Start the titration program.
7. The titrator will adjust the pH of the substrate solution to 7.3 using 0.01 N KOH.
8. Add 50 µl of the sample solution.
9. Allow the titration to proceed automatically. The titrator will record the titration curve and calculate the slope. The slope between 2 and 6 minutes is used by the titrator to calculate the phospholipase C activity. Alternatively, the calculation can be performed manually.

NOTE: The slope must be within 0.02-0.1 ml/min. If the slope is outside this range or if the titration has not started within the first two minutes, adjust the activity of the sample solution.

**Calculation**
Use the following formula for manual calculation of phospholipase C activity:

$$\text{Activity (PLCU/g)} = \frac{V \times DF \times S \times N \times 1000}{V_s \times W}$$

Where:

V is the initial volume of the sample solution (50 ml)

DF is the dilution factor

S is the slope of the titration curve (ml/min)

N is the normality of potassium hydroxide (0.01 mmol/ml)

1000 is the conversion factor from millimoles to micromoles

$V_s$ is the volume of the sample solution used in the assay (0.05 ml)

W is the sample weight (g)

# PHYTOSTEROLS, PHYTOSTANOLS AND THEIR ESTERS

*New specifications prepared at the 69th JECFA (2008), published in FAO JECFA Monographs 5 (2008). An ADI of 0-40 mg/kg bw, expressed as the sum of phytosterols and phytostanols in their free form, was established at the 69th JECFA (2008).*

**SYNONYMS**  Plant sterols/stanols, Plant sterol/stanol esters, Phytosterol/Phytostanol esters

**DEFINITION**  Phytosterols, phytostanols and their esters are a group of steroid alcohols and esters that occur naturally in plants. The B-ring of the steroidal moiety of phytosterols is unsaturated in the 5-6 position and is saturated in phytostanols. Phytosterols and phytostanols are isolated from deoderizer distillate (a by-product of edible oil production), or derived from tall oil (a by-product of wood pulp manufacture). They are purified by distillation, extraction, crystallization and washing resulting in products of high purity. Phytosterol blends derived from either vegetable oils or tall oil may be converted to the corresponding phytostanols by catalytic saturation. Some phytosterols and phytostanols may be extracted as esters of fatty acids. Esters are also produced by reacting the sterol/stanols with fatty acids derived from food grade vegetable oils. The fatty acid ester chain may be saturated, mono- or polyunsaturated depending on the source of the vegetable oil. Commercial products may be mixtures of phytosterols, phytostanols and their esters. The production process may include the use of hexane, 1-propanol, ethanol and methanol.

Chemical names

The major free phytosterols and phytostanols are listed below. In some preparations they are esterified with vegetable oil fatty acids.

Sitosterol: ($3\beta$)-Stigmast-5-en-3-ol
Sitostanol: ($3\beta,5\alpha$)-Stigmastan-3-ol
Campesterol: ($3\beta$)-Ergost-5-en-3-ol
Campestanol: ($3\beta,5\alpha$)-Ergostan-3-ol
Stigmasterol: ($3\beta$)-Stigmasta-5,22-dien-3-ol
Brassicasterol: ($3\beta$)-Ergosta-5,22-dien-3-ol

Esters of sitostanol: for example, sitostanyl oleate
Esters of campesterol: for example, campesteryl oleate

C.A.S numbers

The major free phytosterols and phytostanols are listed below. In some preparations they are esterified with vegetable oil fatty acids. Esterified forms have not been assigned C.A.S numbers

Sitosterol: 83-46-5
Sitostanol: 83-45-4
Campesterol: 474-62-4
Campestanol: 474-60-2
Stigmasterol: 83-48-7
Brassicasterol: 474-67-9

Chemical formula

The major free phytosterols and phytostanols are listed below. In some preparations they are esterified with vegetable oil fatty acids ranging in chain-length from C14 to C18.

Sitosterol: $C_{29}H_{50}O$
Sitostanol: $C_{29}H_{52}O$
Campesterol: $C_{28}H_{48}O$
Campestanol: $C_{28}H_{50}O$
Stigmasterol: $C_{29}H_{48}O$
Brassicasterol: $C_{28}H_{46}O$

Examples of phytosteryl and phytostanyl esters:
Campesteryl oleate: $C_{46}H_{81}O_2$
Sitostanyl oleate: $C_{47}H_{85}O_2$

Structural formulae    Steroid skeleton

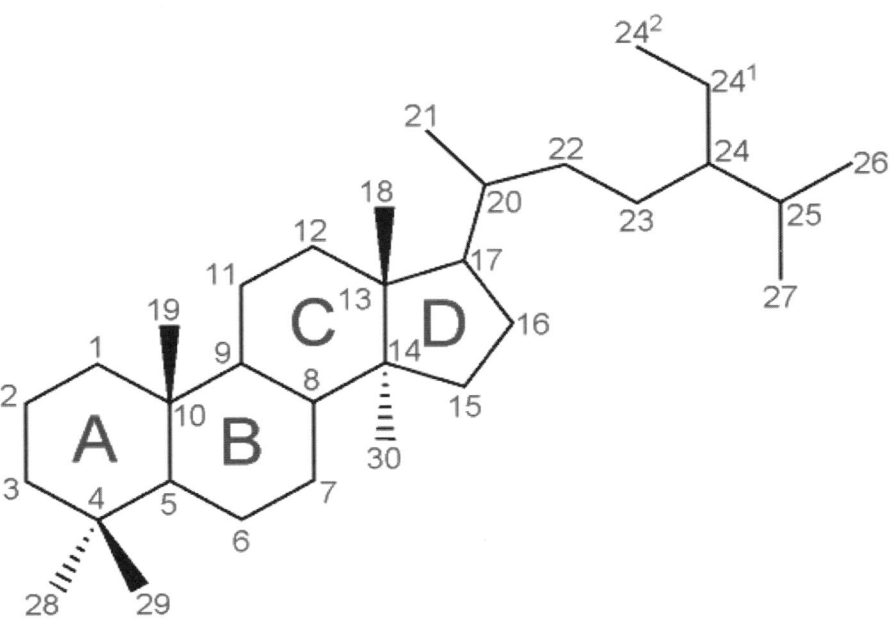

Some examples of phytosterols, phytostanols and a phytostanyl ester

**Sitosterol**

**Campesterol**

**Sitostanol**

**Campestanol**

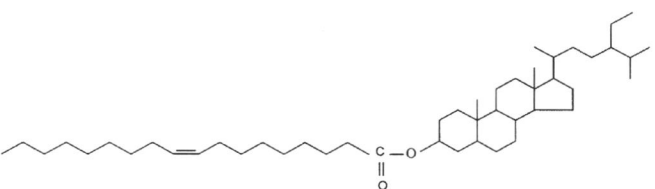
**Sitostanyl oleate**

| Formula weight | Sitosterol: | 414.72 |
|---|---|---|
| | Sitostanol: | 416.73 |
| | Campesterol: | 400.69 |
| | Campestanol: | 402.70 |
| | Stigmasterol: | 412.67 |
| | Brassicasterol: | 398.67 |

Examples of phytosteryl and phytostanyl esters:
Campesteryl oleate: 683.19
Sitostanyl oleate: 699.19

**Assay**	Products containing only free sterols and stanols: not less than 95% on a total free sterol/stanol basis.
Products containing only esterified sterols and stanols: not less than 55% sterol/stanol on a saponifed sample.
Products that are mixtures of free and esterified sterols and stanols: the content of stanols/sterols ranges between 55 and 95% as determined by measurement of free sterols/stanols in a native and saponified sample.
Difference between 55% and 95% is attributable to the fatty acid ester component.

## DESCRIPTION
Free-flowing, white to off-white powders, pills or pastilles; colourless to pale yellow liquids

## FUNCTIONAL USE
This preparation serves no technological purpose in food. It is added to food as a source of phytosterols and phytostanols.

## CHARACTERISTICS

IDENTIFICATION

<u>Solubility</u>	Practically insoluble in water.
Phytosterols and phytostanols are soluble in acetone and ethyl acetate.
Phytosterol and phytostanol esters are soluble in hexane, iso-octane and 2-propanol

<u>Gas Chromatography</u> (Vol. 4)	The retention time for the major peak of a saponified sample in a GC chromatogram of the sample corresponds to that of the β-sitosterol/sitostanol standard using the conditions described in the Method of Assay. The relative retention times of β-sitosterol/sitostanol are approximately 1.066 and 1.073, respectively.

PURITY

<u>Total ash</u> (Vol. 4)	Not more than 0.1 %

<u>Residual solvents</u> (Vol. 4)	Hexane, 1-propanol, ethanol or methanol: 50 mg/kg either singly or

in combination

Water (Vol. 4) — Not more than 4% (Karl Fischer). The selection of sample size and method of sample preparation may be based on the principles of the methods described in Volume 4 (under "General Methods, Water Determination")

Arsenic (Vol. 4) — Not more than 3 mg/kg
Determine by the atomic absorption hydride technique. The selection of sample size and method of sample preparation may be based on the principles of the methods described in Volume 4 (under "General Methods, Metallic Impurities").

Lead (Vol. 4) — Not more than 1 mg/kg
Determine using an AAS/ICP-AES technique appropriate to the specified level. The selection of sample size and method of sample preparation may be based on the principles of the method described in Volume 4 (under "General Methods, Metallic Impurities").

**METHOD OF ASSAY**

Principle
Sterols/stanols are silylated and analysed by gas chromatography with flame ionization detection (Volume 4, "Analytical Techniques, Chromatography"). Esterified sterols/stanols are first saponified and the non-polar components are extracted, dried and silylated. For quantification an internal standard is added to the sample.

Sample preparation
*a. Free sterols/stanols*
Accurately weigh approximately 15 mg 5α-cholestane and approximately 50 mg sterol concentrate into a reaction vial. Add approximately 1 ml methyl tert-butyl ether (MTBE) to dissolve the sample. Warm to 40 – 50° to improve solubility. Add 4.0 ml hexane and mix. Transfer 50 µl of the solution to a small test-tube and evaporate to dryness under nitrogen at 50 – 60°. Add 60 µl N,O-Bis(trimethylsilyl)trifluoroacetamide (BSTFA) and 240 µl pyridine, mix, cap the tube and heat at 60 – 70° for approximately 30 minutes. Mix the solution after 5 – 10 minutes. Add 1.7 ml heptane, mix and transfer the solution to a GC vial.

*b. Sterol/stanol esters*
Accurately weigh approximately 15 mg 5α-cholestane and approximately 100 mg sterol ester accurately into a reaction vial. Add 2 ml ethanolic potassium hydroxide solution (6.6 g KOH in 50 ml ethanol), mix and heat for 90 minutes at 70°. Mix the solution every 15 minutes during saponification. Add 1 ml water and 4 ml heptane to the saponified solution and mix thoroughly for 15 seconds. Wait until the two layers separate completely and transfer the heptane extract to a test-tube. Repeat the extraction twice with 4 ml heptane, collect all three heptane extracts in the same test tube and mix thoroughly. Transfer 50 µl of the solution to a small test-tube and evaporate to dryness under nitrogen at 70 – 80°. Add 60 µl BSTFA and 240 µl pyridine, mix, cap the tube and heat at 60 – 70° for approximately 30 minutes. Mix the solution after 5 – 10 minutes. Add 1.7 ml heptane, mix and transfer the solution to a GC vial.

Equipment
Gas chromatograph, suitable for capillary columns equipped with:

- flame ionization detector (FID)
- cold on-column injector
- autosampler

Capillary column:
- Precolumn: uncoated fused silica capillary, (apolar deactivated), 1.0 m x 0.53 mm i.d. (e.g. Interscience, HRGC precolumn, code 26060370, or equivalent)
- Analytical column 1: CP SIL 13CB, (length 25 m, 0.25 mm i.d.) 0.2 µm film thickness (the dimensions of the column may be altered to accommodate commercially available columns)
- Analytical column 2: CP SIL 8CB, (length 30 m, 0.25 mm i.d.) 0.25 µm film thickness (the dimensions of the column may be altered to accommodate commercially available columns)
All columns are to be connected together with glass quick-seal connectors.

Suitable GC conditions:
- Helium carrier gas flow: 0.9 ml/min
- Detector Temperature: 325°
- FID flow air: 300 ml/min
- FID flow $H_2$: 30 ml/min
- FID flow makeup $N_2$: 30 ml/min

Procedure
Inject 0.5 µl of the sample into the gas chromatograph and run according to the following oven temperature program: 60° (for 1 min), then 15°/min up to 250°, then 2°/min up to 300° (hold for 18 min).

*Peak assignment and identification of individual components*
Identify the main components using a reference sample of known composition. The table of relative retention times given below should be used as a further guide. All other peaks should be identified as unknown.

| Component | Relative retention time (-) |
|---|---|
| 5α-cholestane (internal standard) | 0.761 |
| Cholesterol | 0.929 |
| Cholestanol | 0.934 |
| Brassicasterol | 0.958 |
| Cholestanone | 0.967 |
| 24-methylcholesterol | 0.989 |
| Campesterol | 1.000 |
| Campestanol | 1.007 |
| Stigmasterol | 1.021 |
| Unidentified stanol | 1.028 |
| δ7-campesterol | 1.044 |
| Unidentified sterol 1 | 1.048 |
| Clerosterol | 1.053 |
| Sitosterol | 1.066 |
| Sitostanol | 1.073 |
| δ5-avenasterol | 1.080 |
| Unidentified sterol 2 | 1.094 |
| δ7-stigmastenol | 1.103 |

| | |
|---|---|
| δ7-avenasterol | 1.115 |
| Unidentified sterol 3 | 1.133 |

Calculation of result

Calculation of the concentration of the individual components (mg/kg)

$$C_I = \frac{C_{IS} \times V_{IS} \times A_{component} \times PURITY_{IS} \times 10^6}{A_{IS} \times W_s \times RF}$$

where:
$C_I$ = component
$C_{IS}$ = internal standard concentration (mg/ml)
$V_{IS}$ = internal standard volume (ml)
$A_{component}$ = peak area of individual component
$PURITY_{IS}$ = purity internal standard (%)
$A_{IS}$ = internal standard peak area
$W_s$ = sample weight (mg)
RF = response factor of FID, RF = 1.05 for stanols and 1.00 for other components

Report all sterols/stanols individually. Report the sum of the unidentified sterols/stanols as "unknown sterols/stanols". Report all other peaks in the chromatogram as unknowns (sum value).

# POLYDIMETHYLSILOXANE

*Prepared at the 69th JECFA (2008), published in FAO JECFA Monographs 5 (2008), superseding specifications prepared at the 37th JECFA (1990), published in the Combined Compendium of Food Additive Specifications, FAO JECFA Monographs 1 (2005). A temporary ADI of 0-0.8 mg/kg bw was established at the 69th JECFA (2008).*

| | |
|---|---|
| **SYNONYMS** | Poly(dimethylsiloxane), dimethylpolysiloxane, dimethylsilicone fluid, dimethylsilicone oil; dimethicone; INS No. 900a |
| **DEFINITION** | Polydimethylsiloxane consists of fully methylated linear siloxane polymers containing repeating units of the formula $[(CH_3)_2SiO]$ with trimethylsiloxy end-blocking units of the formula $(CH_3)_3SiO-$. The additive is produced by hydrolysis of a mixture of dimethyldichlorosilane and a small quantity of trimethylchlorosilane. The average molecular weights of the linear polymers range from approximately 6,800 to 30,000. |

(NOTE: In commerce, polydimethylsiloxane is frequently used in preparations usually containing silica gel. The pure substance described in this monograph can be isolated from silica gel-containing liquids by centrifuging at about 20,000 rpm. Before testing the Polydimethylsiloxane for *Identification*, *Refractive index*, *Specific gravity*, and *Viscosity*, any silica gel present must be removed by centrifugation.)

(NOTE: This monograph does not apply to aqueous formulations of Polydimethylsiloxane containing emulsifying agents and preservatives, in addition to silica gel.)

| | |
|---|---|
| Chemical names | α-(Trimethylsilyl)-ω-methylpoly(oxy(dimethylsilylene)) |
| C.A.S. number | 9006-65-9 |
| Structural formula | |

*n* ranges from 90 to 410

| | |
|---|---|
| Assay | Silicon content not less than 37.3% and not more than 38.5% of the total |
| **DESCRIPTION** | Clear, colourless, viscous liquid. |
| **FUNCTIONAL USES** | Antifoaming agent, anticaking agent |
| **CHARACTERISTIS** | |
| IDENTIFICATION | |

| | |
|---|---|
| Solubility (Vol. 4) | Insoluble in water and in ethanol; soluble in most aliphatic and aromatic hydrocarbon solvents |
| Specific gravity (Vol. 4) | $d^{25}_{25}$ : 0.964 - 0.977 |
| Refractive index (Vol. 4) | $n^{25}_D$ : 1.400 - 1.405 |
| Infrared absorption | The infrared absorption spectrum of a liquid film of the sample between two sodium chloride plates exhibits relative maxima at the same wavelengths as those of a similar preparation of USP Dimethylpolysiloxane Reference Standard (available through http://www.usp.org/referenceStandards/catalog.html or by mail to USP 12601 Twinbrook Pkwy, Rockville, MD 20852 USA). |

PURITY

| | |
|---|---|
| Loss on drying (Vol.4) | Not more than 0.5% (150°, 4h) |
| Viscosity | 100 - 1500 cSt at 25°<br>See description under TESTS |
| Lead (Vol. 4) | Not more than 1 mg/kg<br>Determine using an AAS/ICP-AES technique appropriate to the specified level. The selection of sample size and method of sample preparation may be based on principles of methods described in Volume 4 (under "General Methods, Metallic Impurities"). |

## TESTS

PURITY TESTS

| | |
|---|---|
| Viscosity | The Ubbelohde suspended level viscometer, shown in the accompanying diagram, is preferred for the determination of the viscosity. |

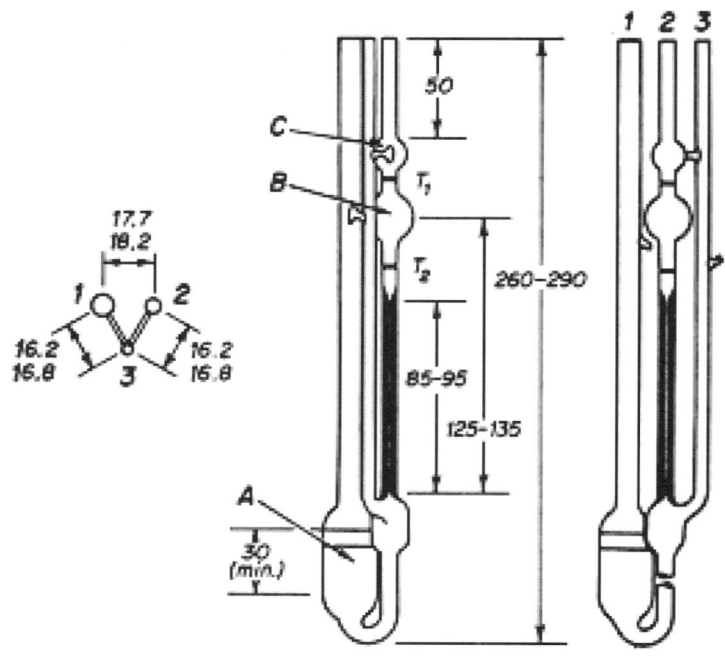

(Dimensions in mm)

For use in the range of 100 to 1,500 centistokes, a No. 3 size viscometer, having a capillary diameter of 2.00 + 0.04 mm, is required. The viscometer should be fitted with holders that satisfy the dimensional positions of the separate tubes as shown in the diagram, and that hold the viscometer vertical. Filling lines in bulb A indicate the minimum and maximum volumes of liquid to be used for convenient operation. The volume of bulb B is approximately 5 ml.

Calibration of the viscometer
Determine the viscosity constant, $k$, for each viscometer by using an oil of known viscosity. [NOTE: Choose an oil with a viscosity as close as possible to that of the sample to be tested.] Charge the viscometer by tilting the instrument about 30 degrees from the vertical, with bulb A below the capillary, and then introduce enough of the sample into tube 1 to bring the level up to the lower filling line. The level should not be above the upper filling line when the viscometer is returned to the vertical position and the sample has drained from tube 1. Charge the viscometer in such a manner that the U-tube at the bottom fills completely without trapping air.

After the viscometer has been in a constant-temperature bath long enough for the sample to reach temperature equilibrium, place a finger over tube 3 and apply suction to tube 2 until the liquid reaches the center of bulb C. Remove suction from tube 2, then remove the finger from tube 3 and place it over tube 2 until the sample drops away from the lower end of the capillary. Remove the finger from tube 2, and measure the time, to the nearest 0.1 sec required for the meniscus to pass from the first time mark ($T_1$) to the second ($T_2$). In order to obtain accurate results within a reasonable time, the apparatus should be adjusted to give an elapsed time of from 80 to 100 sec.

Calculate the viscometer constant $k$ by the equation

$$k = v/t_1,$$

in which $v$ is the viscosity, in centistokes, and $t_1$ is the efflux time, in sec, for the standard liquid.

Viscosity determination of Polydimethylsiloxane
Charge the viscometer with the sample in the same manner as described for the calibration procedure; determine the efflux time, $t_2$; and calculate the viscosity of the sample, $v_s$, by the equation

$$v_s = kt_2.$$

**METHOD OF ASSAY**   Principle
Silicon in the sample is converted to a soluble form by fusion with sodium peroxide. Soluble silicon is measured in the percent range as total silicon by atomic absorption spectrophotometry.

Apparatus
- Fusion apparatus: Parr-type fusion cup; 500-ml nickel beaker; and nickel lid for beaker - or equivalent (avoid use of glassware during fusion and solubilization).
- Instrument: atomic absorption spectrophotometer with silicon hollow cathode lamp; nitrous oxide - acetylene burner, or equivalent.

Reagents
- Sodium peroxide, glacial acetic acid, silica (of known purity for use as standard).

Procedure
[CAUTION: Normal safe laboratory practices for Parr-type bomb fusion should be followed.]

Equivalent fusions must be performed on sample(s), reagent blank(s) and silica standards for each series of samples. For each sample weigh a portion (W) not to exceed 0.3 g into a Parr-type fusion cup (use gelatin capsules for liquid samples). Add 15.0±0.5 g of sodium peroxide.

Assemble the fusion apparatus and place it in a protective ignition rack. Fill the cavity above the cap with water and keep it full during ignition to prevent the gasket from melting. Heat the bottom of the cup with a blast lamp until the cup becomes cherry red about 100 mm up from the bottom within 90 sec. Quench the apparatus in ice water and disassemble the apparatus. Place the cup in the nickel beaker containing 150 to 200 ml of distilled water. Rinse any material adhering to the inside of the assembly cap into the beaker with distilled water. Cover the beaker with the nickel lid. When dissolution is complete and the solution has cooled, remove the cup from the beaker and rinse it with distilled water into the beaker. Add 55.0 ml of reagent grade glacial acetic acid to the beaker. Cool the solution to room temperature and transfer it to a 500 ml volumetric flask. Dilute to volume with distilled water. The solution should contain about 100 µg silicon/ml for a sample weight of about 0.13 g. This method performs best if the silicon concentration of the final analysis solution is 1 to 200 µg/ml. Prepare a series of standards using the same fusion technique that brackets the sample.

Measure the absorbance of sample(s), reagent blank and standards at 251.6 nm with the spectrophotometer according to the manufacturer's operating instructions to obtain optimum analysis conditions for maximum lamp output and fuel and oxidant flow rate to the burner (or equivalent procedures for other vaporizing techniques). Adjust the zero absorbance while aspirating the solvent blank (water) used to dilute the samples. Measure the absorbance of sample(s), reagent blank and standards. Estimate the concentration of silicon in the sample solution from the standards, correcting for the reagent blank. Calculate the percent total silicon in the sample by the equation

$$\%\text{Silicon} = 0.05 \times C/W$$

where
C is the silicon concentration of the sample solution (µg/ml)
W is the weight of sample taken (g)

# STEVIOL GLYCOSIDES

*Prepared at the 69th JECFA (2008), published in FAO JECFA Monographs 5 (2008), superseding specifications prepared at the 68th JECFA (2007), published in FAO JECFA Monographs 5 (2008). An ADI of 0 - 4 mg/kg bw (expressed as steviol) was established at the 69th JECFA (2008).*

**SYNONYMS**  INS no. 960

**DEFINITION**  The product is obtained from the leaves of *Stevia rebaudiana* Bertoni. The leaves are extracted with hot water and the aqueous extract is passed through an adsorption resin to trap and concentrate the component steviol glycosides. The resin is washed with a solvent alcohol to release the glycosides and product is recrystallized from methanol or aqueous ethanol. Ion exchange resins may be used in the purification process. The final product may be spray-dried.

Stevioside and rebaudioside A are the component glycosides of principal interest for their sweetening property. Associated glycosides include rebaudioside C, dulcoside A, rubusoside, steviolbioside, and rebaudioside B generally present in preparations of steviol glycosides at levels lower than stevioside or rebaudioside A.

Chemical name

*Stevioside*: 13-[(2-O-β-D-glucopyranosyl-β-D-glucopyranosyl)oxy] kaur-16-en-18-oic acid, β-D-glucopyranosyl ester

*Rebaudioside A*: 13-[(2-O-β-D-glucopyranosyl-3-O-β-D-glucopyranosyl-β-D-glucopyranosyl)oxy]kaur-6-en-8-oic acid, β-D-glucopyranosyl ester

C.A.S. number

Stevioside: 57817-89-7
Rebaudioside A: 58543-16-1

Chemical formula

Stevioside: $C_{38}H_{60}O_{18}$
Rebaudioside A: $C_{44}H_{70}O_{23}$

| | |
|---|---|
| Structural Formula | The seven named steviol glycosides: |

*[Structure of steviol glycoside aglycone showing CH₃, O-R2, =CH₂, COO-R1 substituents]*

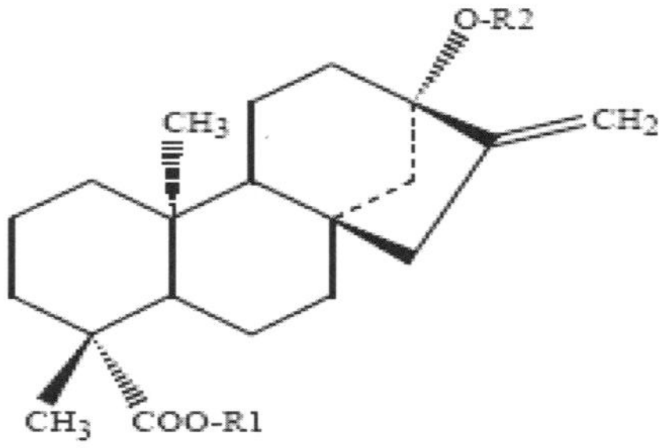

| Compound name | R1 | R2 |
|---|---|---|
| Stevioside | β-Glc | β-Glc-β-Glc(2→1) |
| Rebaudioside A | β-Glc | β-Glc-β-Glc(2→1)<br>\|<br>β-Glc(3→1) |
| Rebaudioside C | β-Glc | β-Glc-α-Rha(2→1)<br>\|<br>β-Glc(3→1) |
| Dulcoside A | β-Glc | β-Glc-α-Rha(2→1) |
| Rubusoside | β-Glc | β-Glc |
| Steviolbioside | H | β-Glc-β-Glc(2→1) |
| Rebaudioside B | H | β-Glc-β-Glc(2→1)<br>\|<br>β-Glc(3→1) |

Steviol (R1 = R2 = H) is the aglycone of the steviol glycosides. Glc and Rha represent, respectively, glucose and rhamnose sugar moieties.

| | |
|---|---|
| Formula weight | Stevioside: 804.88<br>Rebaudioside A: 967.03 |
| Assay | Not less than 95% of the total of the seven named steviol glycosides, on the dried basis. |

## DESCRIPTION

White to light yellow powder, odourless or having a slight characteristic odour. About 200 - 300 times sweeter than sucrose.

## FUNCTIONAL USES

Sweetener

## CHARACTERISTICS

IDENTIFICATION

Solubility (Vol. 4)    Freely soluble in water

| | |
|---|---|
| Stevioside and rebaudioside A | The main peak in the chromatogram obtained by following the procedure in Method of Assay corresponds to either stevioside or rebaudioside A. |
| pH (Vol. 4) | Between 4.5 and 7.0 (1 in 100 solution) |

PURITY

| | |
|---|---|
| Total ash (Vol. 4) | Not more than 1% |
| Loss on drying (Vol. 4) | Not more than 6% (105°, 2h) |
| Residual solvents (Vol. 4) | Not more than 200 mg/kg methanol and not more than 5000 mg/kg ethanol<br>(Method I in Volume 4, General Methods, Organic Components, Residual Solvents) |
| Arsenic (Vol. 4) | Not more than 1 mg/kg<br>Determine by the atomic absorption hydride technique (Use Method II to prepare the test (sample) solution) |
| Lead (Vol. 4) | Not more than 1 mg/kg<br>Determine using an AAS/ICP-AES technique appropriate to the specified level. The selection of sample size and method of sample preparation may be based on the principles of the methods described in Volume 4 (under "General Methods, Metallic Impurities"). |
| **METHOD OF ASSAY** | Determine the percentages of the individual steviol glycosides by high pressure liquid chromatography (Volume 4). |

Standards
Stevioside, >99.0% purity and rebaudioside A, >97% purity (available from Wako pure Chemical Industries, Ltd. Japan).

Mobile phase
Mix HPLC-grade acetonitrile and water (80:20). Adjust the pH to 3.0 with phosphoric acid (85% reagent grade). Filter through 0.22 µm Millipore filter or equivalent.

Standard solutions
(a) Accurately weigh 50 mg of dried (105°, 2 h) stevioside standard into a 100-ml volumetric flask. Dissolve with mobile phase and dilute to volume with mobile phase.
(b) Repeat with previously dried rebaudioside A standard.

Sample solution
Accurately weigh 60-120 mg of dried (105°, 2 h) sample into a 100-ml volumetric flask. Dissolve with mobile phase and dilute to volume with the mobile phase.

Chromatography Conditions
  Column: Supelcosil LC-NH2 or equivalent (length: 15-30 cm; inner diameter: 3.9-4.6 mm)
  Mobile phase: A 80:20 mixture of acetonitrile and water (see above)
  Flow rate: Adjust so that the retention time of rebaudioside A is about 21 min.

Injection volume: 5-10 µl
Detector: UV at 210 nm
Column temperature: 40°

Procedure
Equilibrate the instrument by pumping mobile phase through it until a drift-free baseline is obtained. Record the chromatograms of the sample solution and of the standard solutions.

The retention times relative to rebaudioside A (1.00) are:

0.45-0.48 for stevioside 0.12-0.16 for rubusoside
0.25-0.30 for dulcoside A 0.35-0.41 for steviolbioside
0.63-0.69 for rebaudioside C 0.73-0.79 for rebaudioside B

Measure the peak areas for the seven steviol glycosides from the sample solution (the minor components might not be detected). Measure the peak area for stevioside for the standard solution.

Calculate the percentage of each of the seven steviol glycosides, $X$, in the sample from the formula:

$$\%X = [W_s/W] \times [f_x A_x/A_s] \times 100$$

where
  $W_s$ is the amount (mg) of stevioside in the standard solution
  $W$ is the amount (mg) of sample in the sample solution
  $A_s$ is the peak area for stevioside from the standard solution
  $A_x$ is the peak area of $X$ for the sample solution
  $f_x$ is the ratio of the formula weight of $X$ to the formula weight of stevioside: 1.00 (stevioside), 0.98 (dulcoside A), 1.20 (rebaudioside A), 1.18 (rebaudioside C), 0.80 (rubusoside), 0.80 (steviolbioside), and 1.00 (rebaudioside B).

Calculate the percentage of total steviol glycosides (sum the seven percentages).

# SUNSET YELLOW FCF

*Prepared at the 69th JECFA (2008), published in FAO JECFA Monographs 5 (2008), superseding specifications prepared at the 28th JECFA (1984), published in combined Compendium of Food Additive Specifications, FAO JECFA Monographs 1 (2005). An ADI of 0-2.5 mg/kg bw was established at the 26th JECFA (1982).*

**SYNONYMS** CI Food Yellow 3, Orange Yellow S, CI (1975) No. 15985, INS No. 110

**DEFINITION** Sunset Yellow FCF consists principally of the disodium salt of 6-hydroxy-5-[(4-sulfophenyl)azo]-2-naphthalenesulfonic acid and subsidiary colouring matters together with sodium chloride and/or sodium sulfate as the principal uncoloured components.
(NOTE: The colour may be converted to the corresponding aluminium lake, in which case only the *General Specifications for Aluminium Lakes of Colouring Matters* apply.)

Chemical names
Principal component:
Disodium 6-hydroxy-5-(4-sulfonatophenylazo)-2-naphthalene-sulfonate

C.A.S. number
2783-94-0

Chemical formula
$C_{16}H_{10}N_2Na_2O_7S_2$ (Principal component)

Structural formula

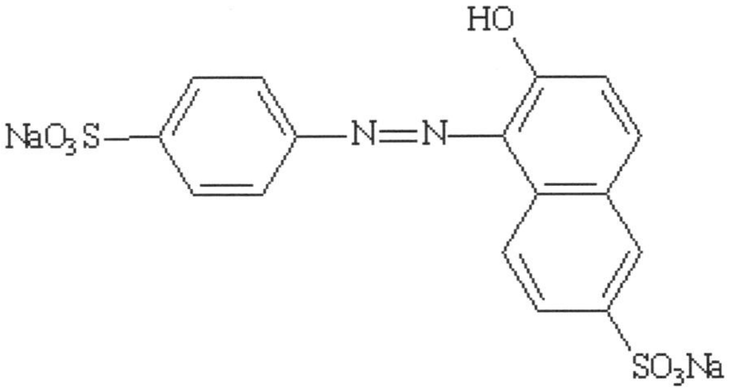

(Principal component)

Formula weight
452.38 (Principal component)

Assay
Not less than 85% total colouring matters

**DESCRIPTION** Orange-red powder or granules

**FUNCTIONAL USES** Colour

**CHARACTERISTICS**

IDENTIFICATION

| | |
|---|---|
| Solubility (Vol. 4) | Soluble in water; sparingly soluble in ethanol |
| Colour test | In water, neutral or acidic solutions of Sunset Yellow FCF are yellow-orange, whereas basic solutions are red-brown. When dissolved in concentrated sulfuric acid, the additive yields an orange solution that turns yellow when diluted with water. |
| Colouring matters, identification (Vol. 4) | Passes test |

PURITY

| | |
|---|---|
| Water content (Loss on drying) (Vol. 4) | Not more than 15% together with chloride and sulfate calculated as sodium salts |
| Water-insoluble matter (Vol. 4) | Not more than 0.2% |
| Lead (Vol. 4) | Not more than 2 mg/kg<br>Determine using an AAS/ICP-AES technique appropriate to the specified level. The selection of sample size and method of sample preparation may be based on the principles of the method described in Volume 4 (under "General Methods, Metallic Impurities"). |
| Subsidiary colouring matter content (Vol. 4) | Not more than 5%<br>Not more than 2% shall be colours other than trisodium 2-hydroxy-1-(4-sulfonatophenylazo)naphthalene-3,6-disulfonate<br>Use the following conditions:<br>Chromatography solvent: 2-Butanone:acetone:water:ammonia (s.g. 0.880) (700:300:300:2)<br>Height of ascent of solvent front: approximately 17 cm |
| Sudan I (1-(Phenylazo)-2-naphthalenol) | Not more than 1 mg/kg<br>See description under TESTS |
| Organic compounds other than colouring matters (Vol. 4) | Not more than 0.5%, sum of the:<br>    monosodium salt of 4-aminobenzenesulfonic acid,<br>    disodium salt of 3-hydroxy-2,7-naphthalenedisulfonic acid,<br>    monosodium salt of 6-hydroxy-2-naphthalenesulfonic acid,<br>    disodium salt of 7-hydroxy-1,3-naphthalenedisulfonic acid,<br>    disodium salt of 4,4'-diazoaminobis-benzenesulfonic acid, and<br>    disodium salt of 6,6'-oxybis-2-naphthalenesulfonic acid<br><br>Proceed as directed under *Determination by High Performance Liquid Chromatography* using an elution gradient of 2 to 100% at 4% per min (linear) followed by elution at 100%. |
| Unsulfonated primary aromatic amines (Vol. 4) | Not more than 0.01%, calculated as aniline |
| Ether-extractable matter (Vol. 4) | Not more than 0.2% |

# TESTS

PURITY TESTS

Sudan I (1-(Phenylazo)-2-naphthalenol)

*Principle*
The additive is dissolved in water and methanol and filtered solutions are analysed by Reverse-Phase Liquid Chromatography (Volume 4 under "Analytical Techniques, Chromatography"), without extraction or concentration. (Based on *J.AOAC Intl* 90, 1373-1378 (2007).)

*Mobile phase*
Eluant A: Ammonium acetate (LC grade), 20 mM aqueous
Eluant B: Methanol (LC grade)

*Sample solution*
Accurately weigh 200 mg of Sunset yellow FCF and transfer it into a 10-ml volumetric flask. Dissolve the sample in 4 ml water via swirling or sonication. Add 5 ml of methanol and swirl. Allow the solution to cool for 5 min and adjust the volume to the mark with water. Filter a part of the solution for analysis through a 13 mm syringe filter with a 0.2 µm pore size PTFE membrane by using a 5 ml polypropylene/polyethylene syringe. (NOTE: Do not substitute a PVDF filter for the PTFE filter, as a PVDF filter adsorbs Sudan I.)

*Standard*
Sudan I (>97%, Sigma Aldrich, or equivalent), recrystallized from absolute ethanol (5g/150 ml)

*Standard stock solution*
Accurately weigh a sufficient quantity of the *Standard* to prepare a solution in methanol of 0.0100 mg/ml.

*Standard solutions*
Transfer 0, 20, 50, 100, 150, 200, and 250 µl aliquots of the *Standard stock solution* to seven 10-ml volumetric flasks. To each flask, add 5 ml of methanol, swirl to mix, and add 4 ml of water. Dilute to volume with water, mix, and filter each solution through a PTFE membrane syringe filter (see *Sample solution*, above) into LC vials for analysis. (NOTE: These solutions nominally contain 0, 0.02, 0.05, 0.10, 0.15, 0.20, and 0.25 µg of Sudan I/ml.)

*Chromatographic system*
Detector: Photodiode Array (485 nm)
Columns: 150 mm x 2.1 mm id, packed with 5 µm reversed-phase C18, or equivalent, with a guard column (10 mm x 2.1 mm i.d.) – Waters Corp., or equivalent
Column temperature: 25°
Flow rate: 0.25 ml/min
Injection volume: 50 µl
Elution: 50% *Eluant A*/50% *Eluant B* for 5 min; 50 to 100% *Eluant B* in 10 min; 100% *Eluant B* for 10 min. (NOTE: The column should be requilibrated with 50% *Eluant A*/50% *Eluant B* for 10 min.)
System suitability: Inject three replicates of the *Standard solutions* with concentrations of 0.05 and 0.25 µg of Sudan I/ml. The responses for each set of three injections show relative standard deviations

of not more than 2%.

*Procedure*
Separately inject the seven *Standard solutions* and the *Sample solution* into the chromatograph and measure the peak areas for Sudan I. From the chromatograms for the *Standard solutions*, prepare a standard curve of the concentration of Sudan I vs the peak areas. (NOTE: The retention time for Sudan I is 19.0 min. Other peaks appearing at earlier retention times in the sample chromatograph are likely attributed to sulfonated subsidiary colours.) Determine the concentration of Sudan I in the *Sample solution* and convert it to mg/kg in the sample of Sunset Yellow FCF.

(NOTE: The limit of determination is 0.4 mg/kg.)

**METHOD OF ASSAY** Proceed as directed under *Colouring Matters Content by Titration with Titanous Chloride* (Volume 4, under "Food Colours, Colouring Matters"), using the following:
    Weight of sample: 0.5-0.6 g
    Buffer: 10 g sodium citrate
    Weight (*D*) of colouring matters equivalent to 1.00 ml of 0.1 N $TiCl_3$: 11.31 mg

# TRISODIUM DIPHOSPHATE

*Prepared at the 69th JECFA (2008), published in FAO JECFA Monographs 5 (2008), based on the previously withdrawn tentative specifications prepared at the 61st JECFA and published in FNP 52, Add 11 (2003). A group MTDI of 70 mg/kg bw, expressed as phosphorus from all food sources, was established at the 26th JECFA (1982).*

| | |
|---|---|
| **SYNONYMS** | Acid trisodium pyrophosphate, trisodium monohydrogen diphosphate; INS No. 450(ii) |
| **DEFINITION** | Trisodium diphosphate is manufactured by calcining sodium orthophosphate having a $Na_2O:P_2O_5$ ratio of 3:2 |
| Chemical names | Trisodium monohydrogen diphosphate |
| C.A.S. number | 14691-80-6 (Anhydrous)<br>26573-04-6 (Monohydrate) |
| Chemical formula | $Na_3HP_2O_7 \cdot x H_2O$ (x = 0 or 1) |
| Formula weight | 243.93 (Anhydrous)<br>261.95 (Monohydrate) |
| Assay | Not less than 57% and not more than 59% expressed as $P_2O_5$ on the dried basis |
| **DESCRIPTION** | White powder or grains |
| **FUNCTIONAL USES** | Stabilizer, leavening agent, emulsifier, nutrient |
| **CHARACTERISTICS** | |
| IDENTIFICATION | |
| Solubility (Vol. 4) | Soluble in water |
| Sodium (Vol. 4) | Passes test |
| Phosphate (Vol. 4) | Passes test |
| PURITY | |
| Loss on drying (Vol. 4) | Anhydrous: Not more than 0.5 % (105°, 4 h)<br>Monohydrate: Not more than 1.0 % (105°, 4 h) |
| Loss on ignition (Vol. 4) | Anhydrous: Not more than 4.5%<br>Monohydrate: Not more than 11.5% |
| Water-insoluble matter (Vol. 4) | Not more than 0.2 % |
| Fluoride (Vol. 4) | Not more than 10 mg/kg<br>See description under TESTS |

Arsenic (Vol. 4)

Not more than 3 mg/kg
Determine by the atomic absorption hydride technique. The selection of sample size and method of sample preparation may be based on the principles of the methods described in Volume 4 (under "General Methods, Metallic Impurities").

Lead (Vol. 4)

Not more than 4 mg/kg
Determine using an atomic absorption/ICP technique appropriate to the specified level. The selection of sample size and method of sample preparation may be based on the principles of the methods described in Volume 4 (under "General Methods, Metallic Impurities").

## TESTS

PURITY TESTS

Fluoride (Vol. 4)

Use Method III. The standard curve constructed in Method III may not be suitable for samples containing low fluoride levels. Therefore, it will be necessary to prepare standard solutions with concentrations other than those specified in Method III for the construction of the standard curve and to choose a sample size that will bring the fluoride concentration within the standard curve.

**METHOD OF ASSAY**  Using a previously dried sample, proceed as directed under *Phosphate Determination as $P_2O_5$, Method I,* Inorganic components (Volume 4). Each ml of 1N sodium hydroxide consumed is equivalent to 3.088 mg of $P_2O_5$ or 5.307 mg of trisodium monohydrogen diphosphate on the dried basis.

# WITHDRAWAL OF SPECIFICATIONS FOR CERTAIN FOOD ADDITIVES

### Carbohydrase from *Aspergillus niger* var.

The Committee reviewed the tentative specifications for carbohydrase from *Aspergillus niger* var. that had been prepared at its $15^{th}$ meeting[1] and for which an ADI "not specified" was established at its $35^{th}$ meeting[2]. The call for data for the $69^{th}$ meeting requested information to revise the existing tentative specifications, stating that the specifications would be withdrawn if no information was forthcoming.

The tentative specifications for carbohydrase include α-amylase, pectinase, cellulase, gluco-amylase, and β-galactosidase (lactase). The functional uses listed in the specifications are diverse and imply that these enzymes are used in food processing as separate enzyme preparations rather than as a mixture of enzymes. Moreover, carbohydrase is not listed as a commercial enzyme by the enzyme industry associations, while all individual enzymes included in the tentative specifications are listed as commercial products.

As no information supporting the tentative specifications was received, the Committee withdrew the ADI and the tentative specifications.

### Estragole

The tentative specifications for estragole used as a food additive that were prepared by the Committee at its $26^{th}$ meeting[3] were withdrawn, as no other uses of estragole other than as a flavouring agent were identified.

---

[1] FAO Nutrition Meeting Series FAO Nutrition Meeting Series, No. 50, 1972 and republished in the Combined Compendium for Food Additive Specifications, FAO JECFA Monographs 1, 2005; WHO Technical Report Series, No. 488, 1972.
[2] WHO Technical Report Series, No. 789, 1990 and corrigenda
[3] FAO Food and Nutrition Paper No. 25, 1982, and republished in the Combined Compendium for Food Additive Specifications, FAO JECFA Monographs 1, 2005

## ANALYTICAL METHODS

The following analytical methods were prepared by the Committee at the 69$^{th}$ meeting. This method will be made available in the on-line edition of Volume 4 of the Combined Compendium of Food Additive Specifications.

# Nickel in Polyols

**Note:** *This method is also applicable for determination of nickel in polydextroses.*

Apparatus
Use a suitable atomic absorption spectrometer equipped with a nickel hollow cathode lamp and an air–acetylene flame to measure the absorbance of the Blank solution, the Standard solutions, and the Sample solution as directed under Procedure (below).

Sample solution
Dissolve 20.0 g of the sample in a mixture of equal volumes of dilute acetic acid TS and water and dilute to 100 ml with the same mixture of solvents. Add 2.0 ml of a 1% w/v solution of ammonium pyrrolidinedithiocarbamate and 10 ml of methyl isobutyl ketone. Mix and allow the layers to separate and use the methylisobutyl ketone layer.

Blank solution
Prepare in the same manner as the Sample solution, but omit the sample.

Standard solutions
Prepare three Standard solutions in the same manner as the Sample solution but adding 0.5 ml, 1.0 ml, and 1.5 ml, respectively, of a standard nickel solution containing 10 mg/kg Ni, in addition to the 20.0 g of the sample.

Procedure
Zero the instrument with the Blank solution. Then determine the absorbances at 232.0 nm of each of the Standard solutions and of the Sample solution at least three times each, and record the average of the steady readings for each. Between each measurement, aspirate the Blank solution, and ascertain that the reading returns to its initial blank value.

Prepare a standard curve by plotting the mean absorbances vs concentration for the Standard solutions. Extrapolate the line joining the points on the graph until it meets the concentration axis. Read the concentration of nickel in the Sample solution at the intersection of the standard curve with the concentration axis.

# SPECIFICATIONS FOR CERTAIN FLAVOURINGS

At its 44th meeting JECFA considered a new approach to the safety evaluation of flavourings. This approach incorporates a series of criteria whose use enables the evaluation of a large number of these agents in a consistent and timely manner. At the $69^{th}$ meeting of the Committee specifications of identity and purity were prepared for 111 new flavourings (page 91).

Information on specifications for flavourings is given on the following tables under the following headings, most of which are self-explanatory:

Name; Chemical name (Systematic name); Synonyms; Flavour and Extract Manufacturers' Association of the United States (FEMA) No; FLAVIS (FL) No; Council of Europe (COE) No; Chemical Abstract Service Registry (CAS) No; Chemical formula (Formula); Molecular weight (M.W.); Physical form/odour; Solubility; Solubility in ethanol, Boiling point (B.P. °C - for information only); Identification test (ID) referring to type of test (NMR: Nuclear Magentic Resonance spectrometry; IR: Infrared spectrometry; MS: Mass spectrometry); Assay min % (Gas chromatographic (GC) assay of flavouring agents); Acid value max; Refractive index (R.I.) (at 20°, if not otherwise stated); Specific gravity (S.G) (at 25°, if not otherwise stated).

The field called "Other requirements" contains four types of entry:

1. Items that are additional requirements, such as further purity criteria or other tests

2. Items provided for information, for example the typical isomer composition of the flavouring agent. These are not considered to be requirements.

3. Substances which are listed as Secondary Constituents (SC) which have been taken into account in the safety evaluation of the named flavouring agent. If the commercial product contains less than 95% of the named compound, it is a requirement that the major part of the product (i.e. not less than 95%) is accounted for by the sum of the named compound and one or more of the secondary constituents.

4. Information on the status of the safety evaluation.

The fields named Session/Status contains the number of the meeting at which the specifications were prepared and the status of the specification. All specifications prepared at the $69^{th}$ meeting were assigned full status.

The the specifications prepared for the 6 alkoxy-substituted allylbenzenes (JECFA Nos 1787 – 1792) by the Committee include a statement that the safety evaluations for these substances had not been completed at the present meeting. For further information see Annex 2.

In addition, the specifications prepared for the group of 40 furan-substituted aliphatic hydrocarbons, alcohols, aldehydes, ketones, carboxylic acids, and related esters, sulfides, disulfides and ethers (JECFA Nos, Structural Class II: 1487, 1488, 1489, 1490, 1491, 1492, 1493, 1494, 1497, 1499, 1503, 1504, 1505, 1507, 1508, 1509, 1510, 1511, 1513, 1514, 1515, 1516, 1517, 1520, 1521, 1522, 1523, 1524, 1525, 1526; Structural Class III: 1495, 1496, 1498, 1500, 1501, 1502, 1506, 1512, 1518, 1519) by the Committee at its $65^{th}$ and $68^{th}$ meetings, will include include a statement that the safety evaluations for these substances had not been completed at the $69^{th}$ meeting. This information is included in the on-line searchable database at the JECFA website at FAO. For further information see Annex 2.

Finally, the reevaluation of the safety of the flavouring substance 2-isopropyl-N,2,3-trimethylbutyramide (JECFA No. 1595) at the $69^{th}$ meeting was not completed due to safety concerns and the specifications in the on-line searchable database at the JECFA website at FAO includes a statement to this effect. For further information see Annex 2.

The spectra used for identification tests are provided from page 108 onwards.

An index listing all the JECFA names is available on page 125.

## NEW SPECIFICATIONS

| No | Name<br>Chemical Name<br>Synonyms | FEMA No<br>FLAVIS No<br>COE No<br>CAS No | Formula<br>M. W. | Physical form;<br>odour | Solubility<br>Solubility in ethanol | B.P. °C | ID test<br>Assay min<br>% | A.V.<br>max | R.I.<br>S. G. | Other requirements/<br>Secondary components | Session<br>Status |
|---|---|---|---|---|---|---|---|---|---|---|---|
| 1787 | Apiole<br>4,7-Dimethoxy-5-(2-propenyl)-1,3-benzodioxole<br>1-Allyl-2,5-dimethoxy-3,4-methylenedioxybenzene | 523-80-8 | C12H14O4<br>222.24 | Colourless to yellow or light green liquid;<br>Slight parsley like aroma | Insoluble in water; soluble in ether, acetone and glacial acetic acid<br>Soluble | 294 | MS<br>95 | | 1.536-1.538<br>1.124-1.135 | Safety evaluation not completed | 69th<br>Full |
| 1788 | Elemicin<br>1,2,3-Trimethoxy-5-(2-propenyl)benzene<br>5-Allyl-1,2,3-trimethoxybenzene | 487-11-6 | C12H16O3<br>208.26 | Colourless to pale straw coloured viscous liquid;<br>Spice with floral notes | Practically insoluble to insoluble in water<br>Soluble | 246 | MS<br>95 | | 1.529-1.534<br>1.058-1.070 | Safety evaluation not completed | 69th<br>Full |
| 1789 | Estragole<br>1-Methoxy-4(2-propenyl)-benzene<br>Methyl chavicol | 2411<br>04.011<br>140-67-0 | C10H12O<br>148.21 | Colourless to light yellow liquid;<br>Anise-like aroma | Insoluble in water; soluble in alcohols<br>Soluble | 216 | IR MS<br>95 | | 1.519-1.524<br>0.960-0.968 | Safety evaluation not completed | 69th<br>Full |
| 1790 | Methyl eugenol<br>1,2-Dimethoxy-4(2-propenyl)-benzene<br>1,2-Dimethoxy-4-allylbenzene | 2475<br>04.012<br>93-15-2 | C11H14O2<br>178.23 | Colourless to pale yellow liquid; Clove-carnation aroma | Insoluble in water; soluble in most fixed oils; insoluble in glycerol and propylene glycol<br>Soluble | 249 | IR MS<br>95 | | 1.532-1.536<br>1.032-1.036 | Safety evaluation not completed | 69th<br>Full |
| 1791 | Myristicin<br>4-Methoxy-6-(2-propenyl)-1,3-benzodioxole<br>5-Allyl-1-methoxy-2,3-(methylenedioxy)benzene | 607-91-0 | C11H12O3<br>192.21 | Colourless oil;<br>Warm balsamic-woody aroma | Practically insoluble to insoluble in water<br>Soluble | 250 | MS<br>95 | | 1.539-1.541<br>1.143-1.145 | Safety evaluation not completed | 69th<br>Full |
| 1792 | Safrole<br>5-(2-Propenyl)-1,3-benzodioxole<br>4-Allyl-1,2-methylenedioxybenzene | 94-59-7 | C10H10O2<br>162.19 | Colourless to slightly yellow liquid;<br>Sasafras aroma | Practically insoluble to insoluble in water<br>Soluble | 232-234 | MS<br>95 | | 1.537-1.540<br>1.095-1.099 | Safety evaluation not completed | 69th<br>Full |
| 1793 | (Z)-2-Penten-1-ol<br>2-Penten-1-ol | 4305<br>02.050<br>665<br>20273-24-9 | C5H10O<br>86.13 | Colourless liquid;<br>Green diffusive aroma | Slightly soluble in water; soluble in non-polar solvents<br>Soluble | 140-141 | MS<br>95 | | 1.427-1.433<br>0.844-0.850 | | 69th<br>Full |

| No | Name<br>Chemical Name<br>*Synonyms* | FEMA No<br>FLAVIS No<br>COE No<br>CAS No | Formula<br>M. W. | Physical form;<br>odour | Solubility<br>Solubility in ethanol | B.P. °C | ID test | Assay min % | A.V. max | R. I.<br>S. G. | Other requirements/<br>Secondary components | Session<br>Status |
|---|---|---|---|---|---|---|---|---|---|---|---|---|
| 1794 | **(E)-2-Decen-1-ol**<br>2-Decen-1-ol | 4304<br>02.137<br>11750<br>18049-18-2 | C10H20O<br>156.27 | Colourless liquid;<br>Fatty rosy aroma | Slightly soluble in water; soluble in non-polar solvents<br>Soluble | 116-118<br>(14 mm Hg) | MS | 95 | | 1.446-1.452<br>0.842-0.848 | | 69th<br>Full |
| 1795 | **(Z)-Pent-2-enyl hexanoate**<br>(Z)-2-Pentenylhexanoic acid ester | 4191<br>09.678<br>74298-89-8 | C11H20O2<br>184.28 | Colourless liquid;<br>banana bergamot aroma | Practically insoluble to insoluble in water; soluble in non-polar solvents<br>Soluble | 240-241 | MS | 95 | | 1.425-1.435<br>0.885-0.895 | | 69th<br>Full |
| 1796 | **(E)-2-Hexenyl octanoate**<br>(E)-2-Hexenyl octanoic acid ester | 4135<br>09.841<br>85554-72-9 | C14H26O2<br>226.36 | Colourless liquid;<br>Pear aroma | Practically insoluble to insoluble in water; soluble in non-polar solvents<br>Soluble | 308-309 | MS | 95 | | 1.448-1.453<br>0.881-0.887 | | 69th<br>Full |
| 1797 | **trans-2-Hexenyl 2-methylbutyrate**<br>(2E)-2-Hexenyl 2-methylbutanoic acid ester | 4274<br>94089-01-7 | C11H20O2<br>184.28 | Liquid;<br>Mild fruity aroma | Insoluble in water; soluble in non-polar solvents<br>Soluble | 231-232 | NMR MS | 95 | 1.0 | 1.430-1.434<br>0.874-0.879<br>(20 °C) | | 69th<br>Full |
| 1798 | **Hept-trans-2-en-1-yl acetate**<br>(2E)-2-Hepten-1-ol acetate | 4125<br>09.385<br>10661<br>16939-73-4 | C9H16O2<br>156.22 | Colourless liquid;<br>Fresh leaf aroma | Practically insoluble to insoluble in water; soluble in non-polar solvents<br>Soluble | 192-193 | MS | 95 | | 1.428-1.434<br>0.889-0.895 | | 69th<br>Full |
| 1799 | **(E,Z)-Hept-2-en-1-yl isovalerate**<br>2-Heptenyl 3-methylbutanoic acid ester | 4126<br>09.303<br>10664<br>253596-70-2 | C12H22O2<br>198.30 | Colourless liquid;<br>Sweet green aroma | Practically insoluble to insoluble in water; soluble in non-polar solvents<br>Soluble | 262-263 | NMR | 95 | | 1.443-1.449<br>0.868-0.873 | | 69th<br>Full |

| No | Name Chemical Name Synonyms | FEMA No FLAVIS No COE No CAS No | Formula M.W. | Physical form; odour | Solubility Solubility in ethanol | B.P. °C | ID test Assay min % | A.V. max | R.I. S.G. | Other requirements/ Secondary components | Session Status |
|---|---|---|---|---|---|---|---|---|---|---|---|
| 1800 | **trans-2-Hexenal glyceryl acetal** | 4273 | C9H16O3 190.24 | Liquid; Weak green and fresh aroma | Slightly soluble in water; soluble in non-polar solvents Soluble | 241-246 | NMR 86 (mixture of isomers) | 1.0 | 1.464-1.474 1.037-1.048 (20 °C) | (+)-2-(1E)-1-Pentenyl-1,3-dioxolane-4-methanol 26%; (+)-2-(1E)-Pentenyl-1,3-dioxan-5-ol 22%; (-)-2-(1E)-Pentenyl-1,3-dioxan-5-ol 22%; (-)-2-(1E)-Pentenyl-1,3-dioxolane-4-methanol 16% SC: 3-Hexenal glyceryl acetal 8%; Hexanal glyceryl acetal 1% | 69th Full |
| | (-)-2-(1E)-Pentenyl-1,3-dioxan-5-ol, (+)-2-(1E)-Pentenyl-1,3-dioxan-5-ol, (-)-2-(1E)-Pentenyl-1,3-dioxolane-4-methanol (+)-2-(1E)-Pentenyl-1,3-dioxolane-4-methanol 2-(1E)-1-Pentenyl-1,3-dioxolane-4-methanol | 214220-85-6/ 897630-96-5/ 897672-50-3/ 897672-51-4 | | | | | | | | | |
| 1801 | **trans-2-Hexenal propylene glycol acetal** | 4272 | C9H16O2 156.22 | Liquid; Weak green and fresh aroma | Slightly soluble in water; soluble in non-polar solvents Soluble | 118-120 (20 mm Hg) | NMR 97 | 1.0 | 1.438-1.444 0.919-0.926 (20 °C) | | 69th Full |
| | 4-Methyl-2-(1E)-1-pentenyl-1,3-dioxolane | 94089-21-1 | | | | | | | | | |
| 1802 | **cis- and trans-1-Methoxy-1-decene** | 4161 | C11H22O 170.29 | Clear, colourless liquid; Fruity floral aroma | Soluble in non-polar solvents; insoluble in water Soluble | 89-90 (9 mm Hg) | NMR IR 98 (Z-isomer 40-48%; E-isomer 52-60%) | | 1.430-1.438 (25 °C) 0.807-0.817 | | 69th Full |
| | 1-Decene, 1-methoxy- 1-Methoxy-1-decene | 79930-37-3 | | | | | | | | | |
| 1803 | **(E)-Tetradec-2-enal** | 4209 05.179 | C14H26O 210.36 | Colourless liquid; Citrus aroma | Practically insoluble to insoluble in water; soluble in non-polar solvents Soluble | 88-89 (0.2 mm Hg) | MS 95 | | 1.455-1.562 0.833-0.841 | | 69th Full |
| | (2E)-Tetradec-2-enal | 51534-36-2 | | | | | | | | | |
| 1804 | **(E)-2-Pentenoic acid** | 4193 08.107 10163 | C5H8O2 100.12 | Colourless liquid; Sour caramellic aroma | Slightly soluble in water; soluble in non-polar solvents Soluble | 197-199 | NMR MS 95 | | 1.445-1.454 0.984-0.991 | | 69th Full |
| | (2E)-2-Pentenoic acid | 13991-37-2 | | | | | | | | | |

| No | Name Chemical Name Synonyms | FEMA No FLAVIS No COE No CAS No | Formula M. W. | Physical form; odour | Solubility Solubility in ethanol | B.P. °C | ID test Assay min % | A.V. max | R. I. S. G. | Other requirements/ Secondary components | Session Status |
|---|---|---|---|---|---|---|---|---|---|---|---|
| 1805 | **(E)-2-Octenoic acid** <br> (2E)-2-Octenoic acid | 3957 <br> 08.114 <br> 10156 <br> 1871-67-6 | C8H14O2 <br> 142.20 | Colourless liquid; Buttery, butterscotch aroma | Insoluble in water; soluble in oils <br><br> Soluble | 139-141 <br> (13 mm Hg) | NMR IR MS <br> 97 | | 1.458-1.462 <br> 0.935-0.941 | | 69th <br> Full |
| 1806 | **Ethyl trans-2-butenoate** <br> 2-Butenoic acid, ethyl ester <br> *Ethyl crotonate* | 3486 <br> 10544-63-5 | C6H10O2 <br> 114.14 | Colourless liquid; Powerful sour caramellic-fruity aroma | Insoluble in water; soluble in oils <br><br> Soluble | 136-137 | MS <br> 98 | 2.0 | 1.422-1.428 <br> 0.916-0.921 | | 69th <br> Full |
| 1807 | **Hexyl 2-butenoate** <br> 2-Butenoic acid, hexyl ester <br> *Hexenyl crotonate* | 3354 <br> 09.266 <br> 10688 <br> 19089-92-0 | C10H18O2 <br> 170.25 | Colourless liquid; Fruity aroma | Insoluble in water, propylene glycol; soluble in most fixed oils <br><br> Soluble | 96-98 <br> (15 mm Hg) | NMR <br> 95 | 1.0 | 1.428-1.442 <br> 0.880-0.905 | | 69th <br> Full |
| 1808 | **Ethyl trans-2-hexenoate** <br> (2E)-2-Hexenoic acid ethyl ester | 3675 <br> 09.850 <br> 631 <br> 27829-72-7 | C8H14O2 <br> 142.20 | Colourless liquid; Fruity, green, pulpy, pineapple, apple aroma | Slightly soluble in water; soluble in fats <br><br> Soluble | 110-111 <br> (10 mm Hg) | NMR IR <br> 95 | | 1.429-1.434 <br> 0.895-0.90 | | 69th <br> Full |
| 1809 | **(E,Z)-Methyl 2-hexenoate** <br> 2-Hexenoic acid, methyl ester <br> *Methyl-beta-propylacrylate* | 2709 <br> 2396-77-2 | C7H12O2 <br> 128.17 | Colourless mobile liquid; Fruity aroma | Very slightly soluble in water; soluble in oils <br><br> Soluble | 168-170 | NMR <br> 95 | | 1.423-1.429 <br> 0.911-0.916 | | 69th <br> Full |
| 1810 | **Hexyl trans-2-hexenoate** <br> Hexyl (E)-2-hexenoate <br> *2-Hexenoic acid hexyl ester* | 3692 <br> 09.292 <br> 33855-57-1 | C12H22O2 <br> 198.31 | Colourless liquid; Fruity, green, slightly fatty aroma | Slightly soluble in water; soluble in fats <br><br> Soluble | 121-123 <br> (10 mm Hg) | NMR IR <br> 92 | | 1.439-1.445 <br> 0.880-0.890 | SC: Hexyl trans-3-hexenoate (6-8%) | 69th <br> Full |
| 1811 | **Methyl trans-2-octenoate** <br> (2E)-2-Octenoic acid methyl ester | 3712 <br> 09.299 <br> 11800 <br> 7367-81-9 | C9H16O2 <br> 156.23 | Colourless liquid; Fruity, green aroma | Slightly soluble in water; soluble in fats <br><br> Soluble | 89-91 <br> (9 mm Hg) | NMR IR MS <br> 90 | | 1.437-1.448 <br> 0.896-0.900 | SC: Methyl trans-3-octenoate (5-6%) | 69th <br> Full |

| No | Name Chemical Name *Synonyms* | FEMA No FLAVIS No COE No CAS No | Formula M.W. | Physical form; odour | Solubility Solubility in ethanol | B.P. °C | ID test Assay min % | A.V. max | R.I. S.G. | Other requirements/ Secondary components | Session Status |
|---|---|---|---|---|---|---|---|---|---|---|---|
| 1812 | Ethyl trans-2-octenoate (2E)-2-Octenoic acid ethyl ester | 3643 09.285 10617 7367-82-0 | C10H18O2 170.24 | Liquid; Green-fruity aroma | Insoluble in water; soluble in fats Soluble | 93-96 (10 mm Hg) | NMR IR MS 98 | | 1.439-1.445 0.888-0.894 (20 °C) | | 69th Full |
| 1813 | (E,Z)-Methyl 2-nonenoate 2-Nonenoic acid, methyl ester *Neofolliione* | 2725 09.234 2099 111-79-5 | C10H18O2 170.25 | Colourless or light-yellow liquid; Green, violet aroma | Insoluble in water; soluble in non-polar solvents Soluble | 114-115 (21 mm Hg) | NMR 95 | 1.0 | 1.440-1.447 0.893-0.900 (20 °C) | | 69th Full |
| 1814 | Ethyl trans-2-decenoate (2E)-2-Decenoic acid, ethyl ester | 3641 09.283 10577 7367-88-6 | C12H22O2 198.31 | Liquid; Fatty-waxy aroma specific to over-ripe pear | Insoluble in water; soluble in fats Soluble | 133-135 (20 mm Hg) | NMR IR MS 95 | 1.0 | 1.440-1.450 0.880-0.890 (20 °C) | | 69th Full |
| 1815 | Ethyl (E)-2-methyl-2-pentenoate 2-Methyl-(2E)-2-pentenoic acid ethyl ester | 4290 1617-40-9 | C8H14O2 142.20 | Clear colourless liquid; Fruity aroma | Slightly soluble in water; soluble in non-polar solvents Soluble | 173-174 | NMR 98 | | 1.436-1.444 0.904-0.914 | | 69th Full |
| 1816 | 2-Methylbutyl 3-methyl-2-butenoate 2-Methylbutyl 3-methyl-2-butenoate *2-Methylbutyl 3-methyl-2-senecioate* | 4306 97890-13-6 | C10H18O2 170.25 | Colourless liquid; Floral fruity aroma | Sparingly soluble in water; soluble in triacetin and propylene glycol Soluble | 57-60 (3.5 mm Hg) | NMR MS 98 | | 1.451-1.461 0.881-0.891 | | 69th Full |
| 1817 | (+/-) (E,Z)-5-(2,2-Dimethylcyclopropyl)-3-methyl-2-pentenal (+/-)(E,Z)-5-(2,2-Dimethylcyclopropyl)-3-methyl-2-pentenal *Acitral* | 4105 877-60-1 | C11H18O 166.27 | Colourless to slightly yellow liquid; Fruity aroma | Insoluble in water Soluble | 234-237 | NMR 90 (E-isomer 45-48% E and Z-isomer 43-45%) | | 1.495-1.501 0.874-0.878 | SC: Citral (<10%) | 69th Full |

| No | Name<br>Chemical Name<br>*Synonyms* | FEMA No<br>FLAVIS No<br>COE No<br>CAS No | Formula<br>M. W. | Physical form;<br>odour | Solubility<br>Solubility in ethanol | B.P. °C | ID test<br>Assay min<br>% | A.V.<br>max | R. I.<br>S. G. | Other requirements/<br>Secondary components | Session<br>Status |
|---|---|---|---|---|---|---|---|---|---|---|---|
| 1818 | (E,Z)-4-Methylpent-2-enoic acid<br>4-Methyl-2-pentenoic acid | 4180<br>08.099<br>10321-71-8 | C6H10O2<br>114.14 | Colourless liquid;<br>Fatty fruity aroma | Slightly soluble in water | 203-204 | MS<br>98 | | 1.442-1.453<br>0.950-0.960 | | 69th<br>Full |
| 1819 | (+/-)-4-Ethyloctanal<br>(+/-)-4-Ethyloctanal | 4117<br>05.223<br>58475-04-0 | C10H20O<br>156.27 | Clear colourless liquid;<br>Floral-like odour | Insoluble in water; soluble in many non-polar solvents<br>Soluble | 97-99<br>(25 mm Hg) | NMR IR MS<br>95 | 1.0 | 1.427-1.434<br>0.834-0.842<br>(20° C) | | 69th<br>Full |
| 1820 | (E)-Geranyl 2-methylbutyrate<br>(2E)-3,7-Dimethyl-2,6-octadienyl 2-methylbutanoic acid | 4122<br>09.382<br>68705-63-5 | C15H26O2<br>238.37 | Colourless liquid;<br>Fruity rosy aroma | Practically insoluble to insoluble in water<br>Soluble | 312-313 | MS<br>95 | | 1.439-1.443<br>0.897-0.903 | | 69th<br>Full |
| 1821 | (E)-Geranyl valerate<br>(2E)-3,7-Dimethyl-2,6-octadienyl pentanoic acid | 4123<br>09.150<br>468<br>1040-247-8 | C15H26O2<br>238.37 | Colourless liquid;<br>Fruity pineapple aroma | Practically insoluble to insoluble in water<br>Soluble | 290-291 | MS<br>95 | | 1.465-1.471<br>0.887-0.900 | | 69th<br>Full |
| 1822 | (E)-Geranyl tiglate<br>2-Methyl- (2E)-2-pentenoic acid ethyl ester<br>*Tiglic acid, geraniol ester* | 4044<br>09.383<br>11829<br>7785-33-3 | C15H24O2<br>236.39 | Very pale yellow liquid; Floral aroma | Insoluble in water<br>Soluble | 271-272 | IR MS<br>96 | 1.0 | 1.477-1.484<br>0.920-0.930<br>(20° C) | | 69th<br>Full |
| 1823 | (E)-Citronellyl 2-methylbut-2-enoate<br>2-Methyl-2-butenoic acid (2E)-3,7-dimethyl-6-octenyl ester | 4295<br>09.340<br>24717-85-9 | C15H26O2<br>238.37 | Colourless liquid;<br>Winey rosy aroma | Practically insoluble to insoluble in water<br>Soluble | 143-145<br>(7 mm Hg) | MS<br>95 | | 1.460-1.470<br>0.901-0.911 | | 69th<br>Full |
| 1824 | (E)-Ethyl tiglate<br>(2E)-2-Methyl- 2-butenoic acid ethyl ester | 2460<br>09.495<br>2185<br>5837-78-5 | C7H12O2<br>128.17 | Colourless liquid;<br>Warm-ethereal fruity aroma | Insoluble in water; soluble in oils<br>Soluble | 154-156 | NMR IR<br>98 | 1.0 | 1.432-1.438<br>0.907-0.916 | | 69th<br>Full |

| No | Name Chemical Name Synonyms | FEMA No FLAVIS No COE No CAS No | Formula M.W. | Physical form; odour | Solubility Solubility in ethanol | B.P. °C | ID test Assay min % | A.V. max | R.I. S.G. | Other requirements/ Secondary components | Session Status |
|---|---|---|---|---|---|---|---|---|---|---|---|
| 1825 | **(E,Z)-Geranic acid** 3,7-Dimethyl-2,6-octadienoic acid | 4121 08.081 10094 459-80-3 | C10H16O2 168.24 | Colourless viscous liquid; Faint floral aroma | Practically insoluble to insoluble in water Soluble | 149-151 (18 mm Hg) | MS 95 (E-isomer 49.4% and Z-isomer 45.6%) | | 1.473-1.479 0.953-0.959 | m.p.= 21 °C | 69th Full |
| 1826 | **Prenyl formate** 3-Methyl-2-buten-1-ol formate | 4205 09.694 68480-28-4 | C6H10O2 114.14 | Colourless liquid; Fruity rum-like aroma | Practically insoluble to insoluble in water Soluble | 34-35 (15 mm Hg) | MS 98 | | 1.410-1.415 0.920-0.927 | | 69th Full |
| 1827 | **Prenyl acetate** 3-Methyl-2-buten-1-ol acetate | 4202 09.692 11796 1191-16-8 | C7H12O2 128.17 | Colourless liquid; Natural green apple banana aroma | Practically insoluble to insoluble in water Soluble | 148-149 | MS 98 | | 1.424-1.428 0.911-0.922 | | 69th Full |
| 1828 | **Prenyl isobutyrate** 2-Methylpropanoic acid 3-methyl-2-butenyl ester | 4206 09.695 76649-23-5 | C9H16O2 156.22 | Colourless liquid; Fruity buttery aroma | Practically insoluble to insoluble in water Soluble | 77-78 (15 mmHg) | MS 99 | | 1.427-1.434 0.887-0.896 | | 69th Full |
| 1829 | **Prenyl caproate** Hexanoic acid 3-methyl-2-butenyl ester | 4204 76649-22-4 | C11H20O2 184.27 | Colourless liquid; Mild green fruit aroma | Practically insoluble to insoluble in water Soluble | 219-221 (25 mm Hg) | MS 96 | | 1.434-1.440 0.880-0.888 | | 69th Full |
| 1830 | **(+/-)-Dihydrofarnesol** 3,7,11-Trimethyl-6,10-dodecadien-1-ol 2,3-*Dihydrofarnesol* | 4031 51411-24-6 | C15H28O 224.39 | Colourless to pale yellow liquid; Floral, fruity aroma | Insoluble in water; soluble in DMSO and acetone Soluble | 301-302 | NMR IR 96 | | 1.471-1.477 0.867-0.873 | | 69th Full |
| 1831 | **(E,Z)-3,7,11-Trimethyldodeca-2,6,10-trienyl acetate** 3,7,11-Trimethyl-2,6,10-dodecatrien-1-ol acetate *Farnesyl acetate* | 4213 09.818 29548-30-9 | C17H28O2 264.41 | Colourless liquid; Rosy floral aroma | Practically insoluble to insoluble in water Soluble | 165-166 (9 mm Hg) | MS 99 (E-isomer 64 %and Z- isomer) 35% | | 1.476-1.479 0.908-0.914 | | 69th Full |

| No | Name / Chemical Name / Synonyms | FEMA No / FLAVIS No / COE No / CAS No | Formula M.W. | Physical form; odour | Solubility / Solubility in ethanol | B.P. °C | ID test / Assay min % | A.V. max | R.I. / S.G. | Other requirements/ Secondary components | Session / Status |
|---|---|---|---|---|---|---|---|---|---|---|---|
| 1832 | **(E,Z)-Phytol**  (2E,7R,11R)-3,7,11,15-Tetramethyl-2-hexadecen-1-ol | 4196  02.204  10302  150-86-7 | C20H40O  296.54 | Colourless to yellow viscous liquid; Faint floral aroma | Practically insoluble to insoluble in water  Soluble | 131-132 (0.1 mm Hg) | MS  95  (E-isomer 65% and Z-isomer 34%) |  | 1.460-1.466  0.847-0.863 |  | 69th  Full |
| 1833 | **(E,Z)-Phytyl acetate**  (2E,7R,11R)-3,7,11,15-Tetramethyl-2-hexadecen-1-ol acetate | 4197  09.691  10236-16-5 | C22H42O2  338.57 | Colourless liquid; Balsamic aroma | Practically insoluble to insoluble in water  Soluble | 129-131 (0.01 mm Hg) | MS  95  (E-isomer 67% and Z-isomer 32%) | 1.0 | 1.451-1.461  0.867-0.873 |  | 69th  Full |
| 1834 | **Methyl 2-methyl-2-propenoate**  2-(Methoxycarbonyl)-1-propene  *Methyl 2-methacrylate* | 4002  09.647  80-62-6 | C5H8O2  100.13 | Clear colourless liquid; Fruity aroma | Slightly soluble in water; soluble in ether and acetone  Soluble | 99-100 | IR  99 |  | 1.411-1.417  0.934-0.938 |  | 69th  Full |
| 1835 | **Isopropenyl acetate**  1-Propen-2-ol acetate | 4152  09.822  108-22-5 | C5H8O2  100.12 | Colourless liquid; Winey ethereal aroma | Practically insoluble to insoluble in water  Soluble | 94-95 | MS  99 |  | 1.397-1.403  0.917-0.923 |  | 69th  Full |
| 1836 | **1-Octen-3-yl acetate**  1-Octen-3-ol acetate | 3582  09.281  11716  2442-10-6 | C10H18O2  170.25 | Almost colourless liquid; Metallic, mushroom aroma | Insoluble in water, propylene glycol; soluble in most fixed oils  Soluble | 189-190 | NMR  95 | 1.0 | 1.420-1.425  0.870-0.876 |  | 69th  Full |
| 1837 | **1-Octen-3-yl butyrate**  Butanoic acid, 1-ethenylhexyl ester  *Butyric acid, 1-pentylallyl ester* | 3612  09.282  16491-54-6 | C12H22O2  198.31 | Colourless liquid; Sweet, fruity, buttery, mushroom aroma | Insoluble in water; soluble in oils; slightly soluble in propylene glycol  Soluble | 80-81 (3.5 mm Hg) | NMR IR MS  95 |  | 1.423-1.428  0.870-0.879 |  | 69th  Full |

| No | Name<br>Chemical Name<br>*Synonyms* | FEMA No<br>FLAVIS No<br>COE No<br>CAS No | Formula<br>M. W. | Physical form;<br>odour | Solubility<br>Solubility in ethanol | B.P. °C | ID test<br>Assay min<br>% | A.V.<br>max | R. I.<br>S. G. | Other requirements/<br>Secondary components | Session<br>Status |
|---|---|---|---|---|---|---|---|---|---|---|---|
| 1838 | **6-Methyl-5-hepten-2-yl acetate**<br>*6-Methyl-5-hepten-2-ol acetate* | 4177<br><br>19162-00-6 | C10H18O2<br>170.25 | Clear colourless liquid; Fruity aroma | Insoluble in water; soluble in non-polar solvents<br>Soluble | 183-184 | NMR IR<br>97 | | 1.420-1.429<br>0.893-0.903 | | 69th<br>Full |
| 1839 | **3-(Hydroxymethyl)-2-octanone**<br>*3-(Hydroxymethyl)-2-octanone* | 3292<br>07.097<br>11113<br>59191-78-5 | C9H18O2<br>158.24 | Colourless oily liquid; Musty, herbaceous, earthy aroma | Slightly soluble in water; soluble in oils<br>Soluble | 78-84<br>(2 mm Hg) | NMR<br>90 | | 1.416-1.422<br>0.874-0.878 | SC: 3-Methylene-2-octanone (7%) | 69th<br>Full |
| 1840 | **(+/-) [R-(E)]-5-Isopropyl-8-methylnona-6,8-dien-2-one**<br>*[R-(E)]-8-Methyl-5-(1-methylethyl)-6,8-nonadien-2-one*<br>*Virginione* | 4331<br>07.239<br><br>2278-53-7 | C13H22O<br>194.35 | Clear yellow liquid; Fruity melon-like aroma | Insoluble in water<br>Soluble | 237-238 | MS<br>95 | | 1.471-1.477<br>0.846-0.852 | | 69th<br>Full |
| 1841 | **(+/-)-cis- and trans-4,8-Dimethyl-3,7-nonadien-2-ol**<br>*4,8-Dimethyl-3,7-nonadien-2-ol* | 4102<br><br>67845-50-5 | C11H20O<br>168.28 | Clear colourless liquid; Green tallowy aroma | Insoluble in water; soluble in most non-polar solvents<br>Soluble | 70-72<br>(2 mm Hg) | NMR IR<br>95 | | 1.465-1.473<br>0.860-0.870 | | 69th<br>Full |
| 1842 | **(+/-)-1-Hepten-3-ol**<br>*(+/-)-1-Hepten-3-ol*<br>*Butyl vinyl carbinol* | 4129<br>02.155<br>10218<br>4938-52-7 | C7H14O<br>114.19 | Colourless liquid; Green strong aroma at high concentration but fatty buttery aroma at low dilution | Insoluble in water; soluble in hexane and diethylether<br>Soluble | 153-154 | NMR IR MS<br>98 | | 1.430-1.437<br>0.831-0.835 | | 69th<br>Full |
| 1843 | **(E,Z)-4-Octen-3-one**<br>*4-Octen-3-one* | 4328<br><br>14129-48-7 | C8H14O<br>126.20 | Clear colourless or pale yellow liquid; Coconut, fruity aroma | Sparingly soluble in water; soluble in many non-polar solvents<br>Soluble | 77-79<br>(20 mm Hg) | NMR<br>95 | | 1.442-1.448<br>0.840-0.844 | | 69th<br>Full |

| No | Name Chemical Name *Synonyms* | FEMA No FLAVIS No COE No CAS No | Formula M. W. | Physical form; odour | Solubility Solubility in ethanol | B.P. °C | ID test Assay min % | A.V. max | R. I. S. G. | Other requirements/ Secondary components | Session Status |
|---|---|---|---|---|---|---|---|---|---|---|---|
| 1844 | **(E)-2-Nonen-4-one** (E)-2-Nonen-4-one | 4301 27743-70-0 | C9H16O 140.22 | Clear colourless or pale yellow liquid; Fruity aroma | Sparingly soluble in water; soluble in non-polar solvents Soluble | 89-91 | NMR 95 | | 1.443-1.449 0.855-0.859 | | 69th Full |
| 1845 | **(E)-5-Nonen-2-one** (E)-5-Nonen-2-one | 4326 27039-84-5 | C9H16O 140.22 | Clear colourless or pale yellow liquid; Fruit reminiscent of berries | Sparingly soluble in water; soluble in many non-polar solvents Soluble | 197-198 | NMR 96 | | 1.433-1.439 0.835-0.839 | | 69th Full |
| 1846 | **(Z)-3-Hexenyl 2-oxopropionate** 3-Oxo-propanoic acid (Z)-3-hexenyl ester | 3934 09.565 10684 68133-76-6 | C9H14O3 170.21 | Colourless liquid; Green, spicy aroma | Insoluble in water; soluble in fats Soluble | 75-77 (5 mm Hg) | NMR IR 98 | | 1.437-1.445 0.982-0.990 | | 69th Full |
| 1847 | **(+/-)-cis and trans-4,8-Dimethyl-3,7-nonadien-2-yl acetate** 4,8-Dimethyl-3,7-nonadien-2-ol acetate | 4103 91418-25-6 | C13H22O2 210.31 | Clear colourless liquid; Green spicy aroma | Insoluble in water; soluble in most non-polar solvents Soluble | 75-83 (2 mm Hg) | NMR IR 95 | | 1.451-1.459 0.890-0.900 | | 69th Full |
| 1848 | **(E)-1,5-Octadien-3-one** 1,5-Octadien-3-one | 4405 07.190 65213-86-7 | C8H12O 124.18 | Colourless liquid; Penetrating earthy aroma | Practically insoluble to insoluble in water Soluble | 168-169 | MS 97 | | 1.424-1.464 0.890-0.900 | | 69th Full |
| 1849 | **10-Undecen-2-one** 10-Undecen-2-one | 4406 36219-73-5 | C11H20O 168.28 | Colourless to pale yellow liquid; Citrus, fatty aroma | Practically insoluble to insoluble in water Soluble | 81-82 (3 mm Hg) | MS 98 | | 1.440-1.441 0.843-0.847 | | 69th Full |
| 1850 | **2,4-Dimethyl-4-nonanol** 2,4-Dimethyl-4-nonanol | 4407 74356-31-3 | C11H24O 172.31 | Colourless liquid; Fruity aroma | Very slightly soluble in water; soluble in fats Soluble | 211-213 | MS 84 | | 1.439-1.447 0.821-0.827 | SC:2,6,8- Trimethyl-6-hydroxy-4-nonanone (6.6%); cis-2,6,8-Trimethyl-5-nonen-4-one (6.5%); trans-2,6,8-Trimethyl-5-nonen-4-one (2.6%) | 69th Full |

| No | Name Chemical Name Synonyms | FEMA No FLAVIS No COE No CAS No | Formula M. W. | Physical form; odour | Solubility Solubility in ethanol | B.P. °C | ID test Assay min % | A.V. max | R. I. S. G. | Other requirements/ Secondary components | Session Status |
|---|---|---|---|---|---|---|---|---|---|---|---|
| 1851 | 8-Nonen-2-one 8-Nonen-2-one | 4408 5009-32-5 | C9H16O 140.22 | Colourless liquid; Fruity aroma | Practically insoluble to insoluble in water Soluble | 91.5-93 (26 mm Hg) | MS 99 | | 1.436-1.437 0.853-0.855 | | 69th Full |
| 1852 | Menthyl valerate 3-Methylbutanoic acid (1R,2S,5R)-5-methyl-2-(1-methylethyl)cyclohexyl ester | 4156 09.154 472 89-47-4 | C15H28O2 240.38 | Colourless liquid; Sweet herbaceous aroma | Practically insoluble to insoluble in water Soluble | 260-262 | NMR MS 95 | | 1.445-1.451 0.903-0.911 | | 69th Full |
| 1853 | 2-(l-Menthoxy)ethanol 2-[[5-Methyl-2-(1-methylethyl)cyclohexyl]oxy]-ethanol 2-(p-Menthan-3-yloxy) ethanol | 4154 38618-23-4 | C12H24O2 200.36 | Clear colourless viscous liquid; Minty aroma | Insoluble in water Soluble | 99-100 (2 mm Hg) | NMR 99 | | 1.457-1.467 0.920-0.940 | | 69th Full |
| 1854 | l-Menthyl acetoacetate 3-Oxobutanoic acid (1R,2S,5R)-5-methyl-2-(1-methylethyl)cyclohexyl ester | 4327 59557-05-0 | C14H24O3 240.34 | Clear colourless or pale yellow liquid; Minty aroma | Sparingly soluble in water; soluble in many non-polar solvents Soluble | 110-115 (2.2 mm Hg) | NMR 96 | | 1.458-1.466 0.979-0.985 | | 69th Full |
| 1855 | l-Menthyl (R,S)-3-hydroxybutyrate 3-Hydroxybutanoic acid 5-methyl-2-(1-methylethyl)cyclohexyl ester | 4308 108766-16-1 | C14H26O3 242.35 | Colourless liquid; Cool minty aroma | Slightly soluble in water; very soluble in corn oil, hexane, ether, chloroform and acetone Soluble | 95-97 (0.5 mm Hg) | NMR IR 95 | 1.0 | 1.454-1.464 0.972-0.985 | | 69th Full |
| 1856 | l-Piperitone (6R)-3-Methyl-6-(1-methylethyl)-2-cyclohexen-1-one | 4200 07.255 2052 4573-50-6 | C10H16O 152.23 | Light yellowish liquid; Herbaceous minty aroma | Insoluble in water Soluble | 233-235 | NMR IR MS 99 | | 1.483-1.487 0.929-0.934 | NOTE: d-isomer is JECFA No. 435 | 69th Full |
| 1857 | 2,6,6-Trimethylcyclohex-2-ene-1,4-dione 2,6,6-Trimethyl-2-cyclohex-2-ene-1,4-dione keto-Isophorone | 3421 07.109 11200 1125-21-9 | C9H12O2 152.20 | White to colourless solid; Woody, musty sweet, aroma | Slightly soluble in water Soluble | NA | NMR IR 98 | | NA NA | m.p. = 23-28 °C | 69th Full |

| No | Name<br>Chemical Name<br>*Synonyms* | FEMA No<br>FLAVIS No<br>COE No<br>CAS No | Formula<br>M. W. | Physical form;<br>odour | Solubility<br>Solubility in ethanol | Solubility in water | B.P. °C | ID test<br>Assay min<br>% | A.V.<br>max | R. I.<br>S. G. | Other requirements/<br>Secondary components | Session<br>Status |
|---|---|---|---|---|---|---|---|---|---|---|---|---|
| 1858 | **Menthyl pyrrolidone carboxylate**<br>2-Isopropyl-5-methyl cyclohexyl 5-oxo-2-pyrrolidine carboxylate<br>*D- and L-Proline* | 4155<br><br>68127-22-0 | C15H25NO3<br>267.36 | Agglomerated fine white powder; Cool refreshing aroma | Slightly soluble in water<br>Soluble | | NA | NMR IR MS<br>98 | | NA<br>NA | m.p. = 68-72 °C | 69th<br>Full |
| 1859 | **3,9-Dimethyl-6-(1-methylethyl)-1,4-dioxaspiro[4.5]decan-2-one**<br>3,9-Dimethyl-6-(1-methylethyl)-1,4-dioxaspiro[4.5]decan-2-one<br>*Freshone* | 4285<br>06.136<br>831213-72-0 | C13H22O3<br>226.30 | Colourless liquid;<br>Minty aroma | Slightly soluble in water; soluble in fats<br>Soluble | | 323-325 | NMR IR MS<br>98 | | 1.458-1.461<br>1.018-1.021 | | 69th<br>Full |
| 1860 | **8-p-Menthene-1,2-diol**<br>1-Methyl-4-(1-methylethenyl)-1,2-cyclohexanediol<br>*Limonene glycol* | 4409<br><br>1946-00-5 | C10H18O2<br>170.25 | Colourless to very slightly yellow oily liquid; Cool minty aroma | Slightly soluble in water<br>Soluble | | 54-57 | MS<br>98 | | 1.493-1.499<br>0.920-0.925 | | 69th<br>Full |
| 1861 | **d-2,8-p-Menthadien-1-ol**<br>1-Methyl-4-(1-methylethenyl)-2-cyclohexen-1-ol | 4411<br><br>22771-44-4 | C10H16O<br>153.23 | Colourless to very slightly yellow oily liquid; Terpiniod aroma | Sparingly soluble in water<br>Soluble | | 247-251 | MS<br>95 | | 1.484-1.494<br>0.936-0.946<br>(20 °C) | | 69th<br>Full |
| 1862 | **Dehydronootkatone**<br>[4R-(4alpha,4a alpha,6beta]-4,4a,5,6-Tetrahydro-4,4a-dimethyl-6-(1-methylethenyl)-2(3H)-naphthalenone<br>*8,9-Didehydronootkatone* | 4091<br><br>5090-63-1 | C15H20O<br>216.33 | Pale yellow to brown liquid; Fruity aroma with citrus undertone | Practically insoluble or insoluble in water; soluble in non-polar solvents<br>Insoluble | | 129-130 | NMR<br>95 | | 1.559-1.569<br>1.009-1.019 | | 69th<br>Full |
| 1863 | **Isobornyl isobutyrate**<br>2-Methylpropanoic acid (1R,2R,4R)-1,7,7-trimethylbicyclo[2.2.1]hept-2-yl ester | 4146<br>09.584<br>85586-67-0 | C14H24O2<br>224.34 | Colourless liquid;<br>Earthy camphorous aroma | Practically insoluble to insoluble in water<br>Soluble | | 131-133<br>(19 mm Hg) | MS<br>95 | | 1.460-1.466<br>0.958-0.964 | | 69th<br>Full |
| 1864 | **l-Bornyl acetate**<br>(1S,2R,4S)-1,7,7-Trimethylbicyclo[2.2.1]heptan-2-ol acetate | 4080<br>09.848<br>5655-61-8 | C12H20O2<br>196.29 | Colourless solid;<br>Sweet herbaceous odour | Practically insoluble to insoluble in water<br>Soluble | | 224-226 | NMR<br>95 | | 1.456-1.462<br>0.981-0.987 | m.p. = 29 °C | 69th<br>Full |

| No | Name Chemical Name Synonyms | FEMA No FLAVIS No COE No CAS No | Formula M. W. | Physical form; odour | Solubility Solubility in ethanol | B.P. °C | ID test Assay min % | A.V. max | R. I. S. G. | Other requirements/ Secondary components | Session Status |
|---|---|---|---|---|---|---|---|---|---|---|---|
| 1865 | **Thujyl alcohol** (1S,3S,4R,5R)-4-Methyl-1-(1-methylethyl)-bicyclo[3.1.0]hexan-3-ol *(-)-3-Neoisothujanol* | 4079 02.207 21653-20-3 | C10H18O 154.25 | Colourless crystals; Minty camphorous odour | Practically insoluble in water; soluble in non-polar solvents Soluble | 99-100 (12 mm Hg) | NMR 95 | | 1.460-1.466 0.919-0.925 | m.p. = 28 °C | 69th Full |
| 1866 | **Vetiverol** 1,2,3,3a,4,5,6,8a-Octahydro-4,8-dimethyl-2-(1-methylethylidene)-6-azulenol | 4217 02.214 10321 89-88-3 | C15H24O 220.35 | Amber solid; Sweet balsamic aroma | Practically insoluble in water Soluble | NA | NMR 95 | | NA NA | m.p. = 69-71 °C | 69th Full |
| 1867 | **Vetiveryl acetate** 1,2,3,3a,4,5,6,8a-Octahydro-4,8-dimethyl-2-(1-methylethylidene)-6-azulenol acetate | 4218 09.821 11887 117-98-6 | C17H26O2 262.39 | Colourless solid; Sweet woody aroma | Practically insoluble in water Soluble | NA | NMR 95 | | NA NA | m.p. = 73 °C | 69th Full |
| 1868 | **3-Pinanone** 2,6,6-Trimethylbicyclo[3.1.1]heptan-3-one *Isopinocamphone* | 4198 07.171 11125 18358-53-7 | C10H16O 154.24 | Colourless liquid; Cedar camphor aroma | Practically insoluble in water Soluble | 69-71 (5 mm Hg) | NMR MS 95 | | 1.472-1.478 0.963-0.969 | | 69th Full |
| 1869 | **Isobornyl 2-methylbutyrate** 2-Methylbutanoic acid 1,7,7-trimethylbicyclo[2.2.1]hept-2-yl ester | 4147 09.888 94200-10-9 | C15H26O2 238.37 | Colourless solid; Herbaceous woody aroma | Practically insoluble in water Soluble | NA | NMR MS 95 | | NA NA | m.p. = 81-84 °C | 69th Full |
| 1870 | **Verbenone** 4,6,6-Trimethylbicyclo[3.1.1]heptan-3-one *Pin-2-en-4-one* | 4216 07.196 11186 80-57-9 | C10H14O 150.22 | Colourless liquid; Minty spicy aroma | Practically insoluble in water; soluble in non-polar solvents Soluble | 89-90 (12 mm Hg) | NMR MS 95 | | 1.490-1.500 0.975-0.981 | | 69th Full |

| No | Name Chemical Name *Synonyms* | FEMA No FLAVIS No COE No CAS No | Formula M. W. | Physical form; odour | Solubility Solubility in ethanol | B.P. °C | ID test Assay min % | A.V. max | R. I. S. G. | Other requirements/ Secondary components | Session Status |
|---|---|---|---|---|---|---|---|---|---|---|---|
| 1871 | **Methyl hexanoate** Methyl hexanoate | 2708 09.069 319 106-70-7 | $C_7H_{14}O_2$ 130.18 | Colourless to pale yellow liquid; Pineapple, ethereal aroma | Insoluble in water; soluble in propylene glycol and vegetable oils Soluble | 150-151 | NMR 98 | 1.0 | 1.402-1.409 0.880-0.889 | | 69th Full |
| 1872 | **Hexyl heptanoate** 1-Hexyl heptanoate | 4337 1119-06-8 | $C_{13}H_{26}O_2$ 214.32 | Liquid; Herbaceous aroma | Insoluble in water; soluble in non-polar solvents Soluble | 252-253 | MS 98 | 1.0 | 1.426-1.430 0.860-0.865 (20 °C) | | 69th Full |
| 1873 | **Hexyl nonanoate** Hexyl nonanoate | 4339 6561-39-3 | $C_{15}H_{30}O_2$ 242.40 | Liquid; Fresh vegetable fruity aroma | Insoluble in water; soluble in non-polar solvents Soluble | 291-292 | MS 96 | 1.0 | 1.431-1.436 0.858-0.863 (20 °C) | | 69th Full |
| 1874 | **Hexyl decanoate** Hexyl decanoate *Hexyl caprate* | 4342 10448-26-7 | $C_{16}H_{32}O_2$ 256.42 | Liquid; Fresh green aroma | Insoluble in water; soluble in non-polar solvents Soluble | 305-306 | MS 98 | 1.0 | 1.432-1.438 0.857-0.863 (20 °C) | | 69th Full |
| 1875 | **Heptyl heptanoate** 1-Heptyl heptanoate | 4341 624-09-9 | $C_{14}H_{28}O_2$ 228.37 | Liquid; Green aroma | Insoluble in water, soluble in non-polar solvents Soluble | 276-277 | MS 98 | 1.0 | 1.428-1.432 0.859-0.865 (20 °C) | | 69th Full |
| 1876 | **Dodecyl propionate** Dodecyl propionate | 4338 6221-93-8 | $C_{15}H_{30}O_2$ 242.40 | Liquid; Slightly fruity light aroma | Insoluble in water; soluble in non-polar solvents Soluble | 283-284 | NMR IR MS 98 | 1.0 | 1.432-1.436 0.860-0.866 (20 °C) | | 69th Full |
| 1877 | **Dodecyl butyrate** Dodecyl butyrate | 4340 3724-61-6 | $C_{16}H_{32}O_2$ 256.42 | Liquid; Slightly fruity light aroma | Insoluble in water; soluble in non-polar solvents Soluble | 305-306 | NMR MS 98 | 1.0 | 1.433-1.438 0.857-0.862 (20 °C) | | 69th Full |

| No | Name Chemical Name Synonyms | FEMA No FLAVIS No COE No CAS No | Formula M. W. | Physical form; odour | Solubility Solubility in ethanol | B.P. °C | ID test Assay min % | A.V. max | R. I. S. G. | Other requirements/ Secondary components | Session Status |
|---|---|---|---|---|---|---|---|---|---|---|---|
| 1878 | **4-Hydroxy-3,5-dimethoxy benzaldehyde** 4-Hydroxy-3,5-dimethoxybenzaldehyde *Galladehyde 3,5-dimethyl ether* | 4049 05.153 10340 134-96-3 | C9H10O4 182.18 | Very pale green needles; Alcoholic aroma | Insoluble in water Soluble | NA | NMR IR 98 | | NA NA | m.p. = 110-113 °C | 69th Full |
| 1879 | **Vanillin 3-(l-menthoxy)propane-1,2-diol acetal** 2-Methoxy-4-[[[5-methyl-2-(1-methylethyl)cyclohexyl]oxy]methyl]-1,3-dioxolan-2-yl]-phenol | 3904 02.248 180964-47-0 | C21H32O5 364.48 | Colourless powder; Minty aroma with vanilla undertones | Slightly soluble in water; soluble in fats, non-polar solvents and acetone Soluble | NA | NMR MS 94 | | NA NA | m.p. = 78-80 °C SC: Vanillin (2-3%) | 69th Full |
| 1880 | **Sodium 4-methoxybenzoyloxyacetate** Benzoic acid, 4-methoxy-, carboxymethyl ester, sodium salt | 4016b 17114-82-8 | C10H9O5Na 232.17 | White solid; Cooked brown and roasted aroma | Slightly soluble in water; insolube in n-octane Soluble | NA | NMR 98 | | NA NA | m.p. = 135 °C | 69th Full |
| 1881 | **Divanillin** 6,6'-Dihydroxy-5,5'-dimethoxy-[1,1'-biphenyl]-3,3'-dicarboxaldehyde *Dehydrodivanillin* | 4107 05.221 2092-49-1 | C16H14O6 302.28 | White solid; Fruity vanilla aroma | Practically insoluble to insoluble in water; soluble in benzyl alcohol Soluble | NA | NMR 91 | | NA NA | m.p. = 315 °C SC: Vanillin (5-7%) | 69th Full |
| 1882 | **Vanillin propylene glycol acetal** 2-Methoxy-4-(4-methyl-1,3-dioxalan-2-yl)-phenol | 3905 06.104 68527-74-2 | C11H14O4 210.23 | Colourless, viscous liquid; Sweet, vanilla aroma | Insoluble in water and fat; soluble in triacetin Soluble | 152-155 (1 mm Hg) | NMR 79 | | 1.533-1.543 1.196-1.206 | SC: Vanillin (18-20%) | 69th Full |
| 1883 | **4-Methoxybenzoyloxyacetic acid** Benzoic acid, 4-methoxy-, carboxymethyl ester *Glycolic acid, p-anisate* | 4016 10414-68-3 | C10H10O5 210.18 | White solid; Cooked brown and roasted aroma | Slightly soluble in water; insoluble in n-octane Soluble | NA | NMR IR MS 98 | | NA NA | m.p. = 135 °C | 69th Full |
| 1884 | **Methyl isothiocyanate** Isothiocyanatomethane | 4426 556-61-6 | C2H3NS 73.11 | Colourless to tan liquid; Pungent, penetrating mustard-like odour | Very slightly soluble in water; freely soluble in ether Soluble | 117-118 | MS 96 | | 1.495-1.499 0.938-0.942 | | 69th Full |

| No | Name<br>Chemical Name<br>*Synonyms* | FEMA No<br>FLAVIS No<br>COE No<br>CAS No | Formula<br>M. W. | Physical form;<br>odour | Solubility<br>Solubility in ethanol | B.P. °C | ID test<br>Assay min<br>% | A.V.<br>max | R. I.<br>S. G. | Other requirements/<br>Secondary<br>components | Session<br>Status |
|---|---|---|---|---|---|---|---|---|---|---|---|
| 1885 | **Ethyl isothiocyanate**<br>Isothiocyanatoethane | 4420<br><br>542-85-8 | C3H5NS<br>87.14 | Colourless liquid;<br>Sharp mustard-like<br>aroma | Very slightly soluble in water; freely soluble in ether<br>Soluble | 130-132 | MS<br>99 | | 1.510-1.515<br>0.997-1.004 | | 69th<br>Full |
| 1886 | **Isobutyl isothiocyanate**<br>1-Isothiocyanato-2-methylpropane | 4424<br><br>591-82-2 | C5H9NS<br>115.20 | Colourless to yellow liquid; Green pungent aroma | Very slightly soluble in water; freely soluble in ether<br>Soluble | 72-73<br>(30 mm Hg) | MS<br>97 | | 1.491-1.499<br>0.935-0.945 | | 69th<br>Full |
| 1887 | **Isoamyl isothiocyanate**<br>1-Isothiocyanato-3-methylbutane | 4423<br><br>628-03-5 | C6H11NS<br>129.23 | Colourless to yellow liquid; Sharp green irritating aroma | Very slightly soluble in water; freely soluble in ether<br>Soluble | 80-82<br>(12 mm Hg) | MS<br>98 | | 1.493-1.499<br>0.939-0.945 | | 69th<br>Full |
| 1888 | **Isopropyl isothiocyanate**<br>2-Isothiocyanatopropane | 4425<br><br>2253-73-8 | C4H7NS<br>101.17 | Colourless liquid; Penetrating mustard-like aroma | Very slightly soluble in water; freely soluble in ether<br>Soluble | 68-70<br>(67 mm Hg) | MS<br>95 | | 1.489-1.497<br>0.947-0.955 | | 69th<br>Full |
| 1889 | **3-Butenyl isothiocyanate**<br>4-Isothiocyanato-1-butene | 4418<br>12.283<br>3386-97-8 | C5H7NS<br>113.20 | Colourless liquid; Penetrating aroma | Very slightly soluble in water; freely soluble in ether<br>Soluble | 75-77<br>(14 mm Hg) | MS<br>97 | | 1.520-1.526<br>0.990-0.996<br>(20 °C) | | 69th<br>Full |
| 1890 | **2-Butyl isothiocyanate**<br>2-Isothiocyanatobutane | 4419<br><br>4426-79-3 | C5H9NS<br>115.20 | Colourless to yellow liquid; Sharp green slightly irritating aroma | Very slightly soluble in water; freely soluble in ether<br>Soluble | 69-70<br>(27 mm Hg) | MS<br>97 | | 1.490-1.497<br>0.938-0.946 | | 69th<br>Full |
| 1891 | **Amyl isothiocyanate**<br>1-Isothiocyanatopentane | 4417<br><br>629-12-9 | C6H11NS<br>129.23 | Colourless to yellow liquid; Sharp green irritating aroma | Very slightly soluble in water; freely soluble in ether<br>Soluble | 101-103<br>(35 mm Hg) | MS<br>97 | | 1.495-1.501<br>0.942-0948<br>(20 °C) | | 69th<br>Full |

| No | Name Chemical Name *Synonyms* | FEMA No FLAVIS No COE No CAS No | Formula M. W. | Physical form; odour | Solubility Solubility in ethanol | B.P. °C | ID test Assay min % | A.V. max | R. I. S. G. | Other requirements/ Secondary components | Session Status |
|---|---|---|---|---|---|---|---|---|---|---|---|
| 1892 | 4-(Methylthio)butyl isothiocyanate *1-Isothiocyanato-4-(methylthio)butane* | 4414 4430-36-8 | C6H11NS2 161.29 | Pale yellow liquid; Penetrating reddish-like aroma | Very slightly soluble in water; freely soluble in ether Soluble | 134-136 (14 mm Hg) | MS 99 | | 1.551-1.556 1.080-1.086 (20 °C) | | 69th Full |
| 1893 | 4-Pentenyl isothiocyanate *5-Isothiocyanato-1-pentene* | 4427 18060-79-2 | C6H9NS 127.21 | Colourless to pale yellow liquid; Strong pungent irritating aroma | Very slightly soluble in water; freely soluble in ether Soluble | 57-58 (3 mm Hg) | MS 95 | | 1.513-1.519 0.970-0.976 (20 °C) | | 69th Full |
| 1894 | 5-Hexenyl isothiocyanate *6-Isothiocyanato-1-hexene* | 4421 49776-81-0 | C7H11NS 141.24 | Colourless to pale yellow liquid; Pungent irritating aroma | Very slightly soluble in water; freely soluble in ether Soluble | 74-76 (3 mm Hg) | MS 96 | | 1.506-1.516 0.955-0.965 (20 °C) | | 69th Full |
| 1895 | Hexyl isothiocyanate *1-Isothiocyanatohexane* | 4422 4404-45-9 | C7H13NS 143.25 | Colourless to yellow liquid; sharp green irritating aroma | Very slightly soluble in water; freely soluble in ether Soluble | 72-73 (8 mm Hg) | MS 97 | | 1.490-1.494 0.931-0.941 (20 °C) | | 69th Full |
| 1896 | 5-(Methylthio)pentyl isothiocyanate *1-Isothiocyanato-5-(methylthio)-pentane* | 4416 4430-42-6 | C7H13NS2 175.32 | Pale yellow liquid; Penetrating reddish-like aroma | Very slightly soluble in water; freely soluble in ether Soluble | 131-133 (4 mm Hg) | MS 96 | | 1.542-1.548 1.055-1.061 (20 °C) | | 69th Full |
| 1897 | 6-(Methylthio)hexyl isothiocyanate *1-Isothiocyanato-6-(methylthio)-hexane* | 4415 4430-39-1 | C8H15NS2 189.34 | Pale yellow liquid; Penetrating reddish-like aroma | Very slightly soluble in water; freely soluble in ether Soluble | 128-129 (1 mm Hg) | MS 95 | | 1.534-1.540 1.035-1.041 (20 °C) | | 69th Full |

## Spectra of certain flavourings

### 1787  Apiole (MS)
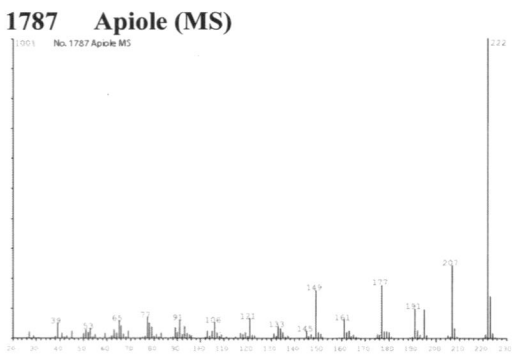

### 1788  Elemicin (MS)
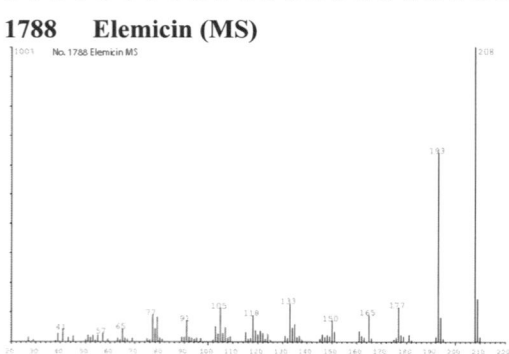

### 1789  Estragole (IR)
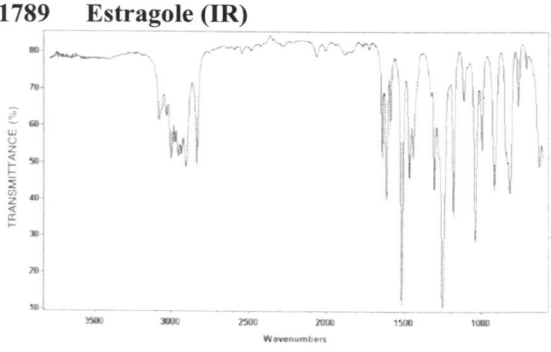

### 1789  Estragole (MS)
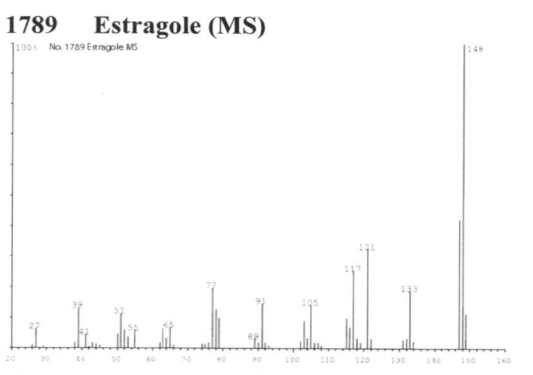

### 1790  Methyl eugenol (IR)
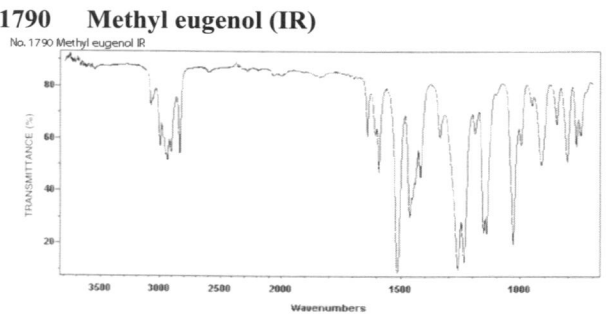

### 1790  Methyl eugenol (MS)
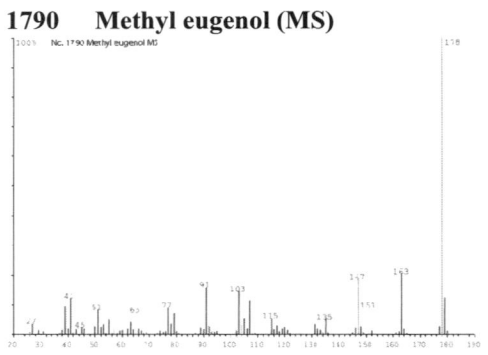

### 1791  Myristicin (MS)
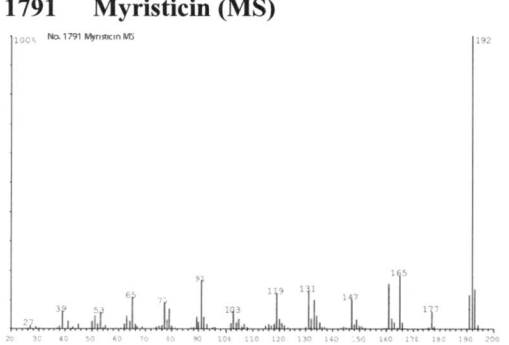

### 1792  Safrole (MS)
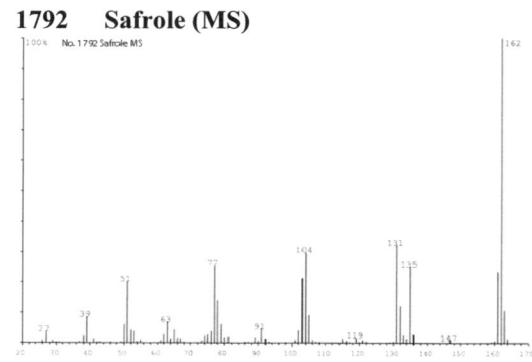

### 1793  (Z)-2-Penten-1-ol (MS)
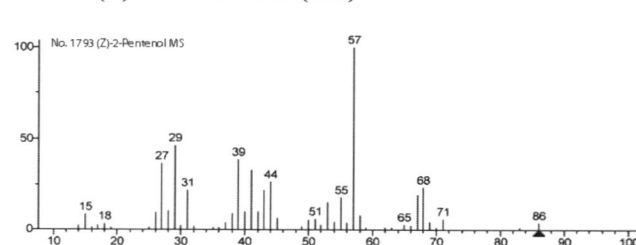
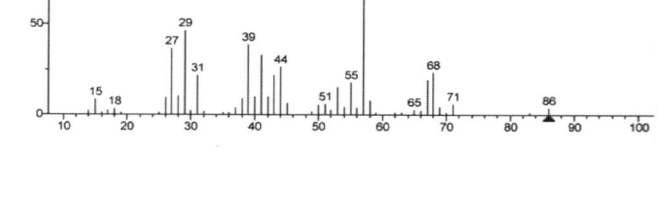

### 1794  (E)-2-Decen-1-ol (MS)
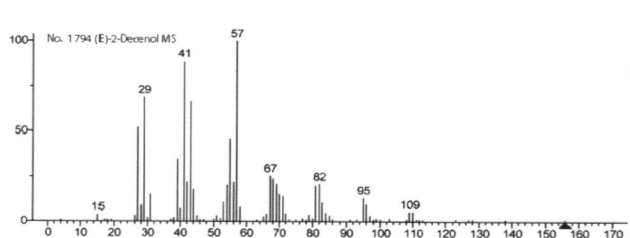

**1795   (Z)-Pent-2-enyl hexanoate (MS)**

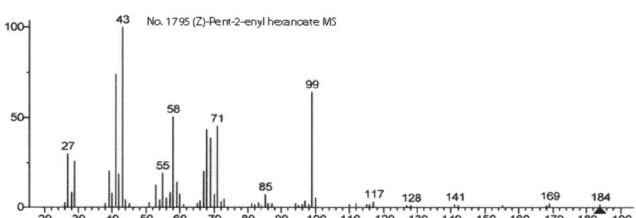

**1796   (E)-2-Hexenyl octanoate (MS)**

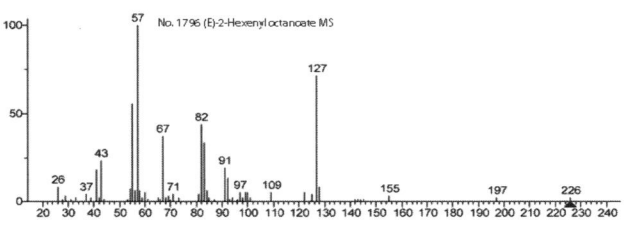

**1797   trans-2-Hexenyl 2-methylbutyrate (1H-NMR)**

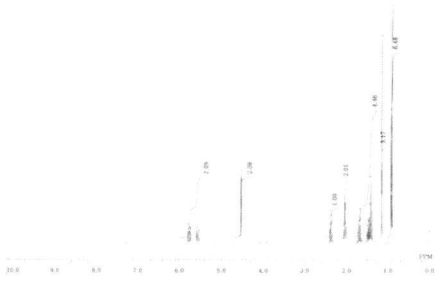

**1797   trans-2-Hexenyl 2-methylbutyrate (MS)**

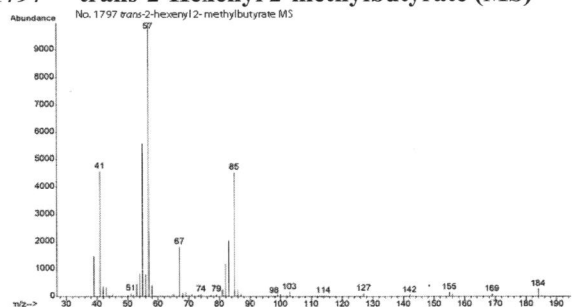

**1798   Hept-trans-2-en-1-yl acetate (MS)**

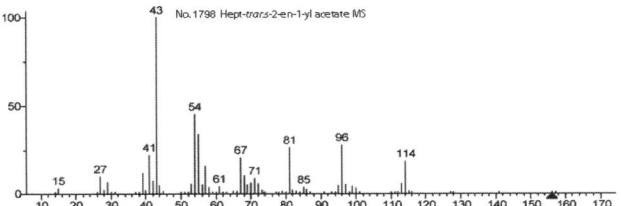

**1799   (E,Z)-Hept-2-en-1-yl isovalerate (13C-NMR)**

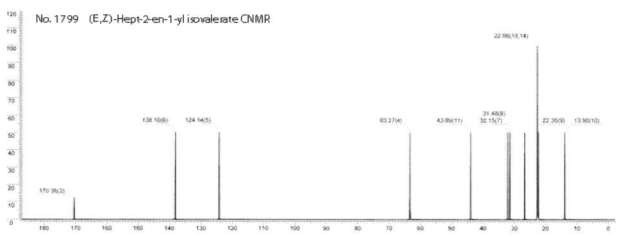

**1800   trans-2-Hexenal glyceryl acetal (1H-NMR)**

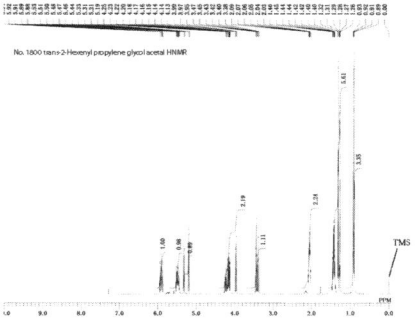

**1801   trans-2-Hexenal propylene glycol acetal (1H-NMR)**

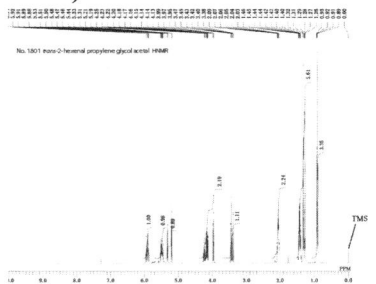

**1802   cis- and trans-1-Methoxy-1-decene (1H-NMR)**

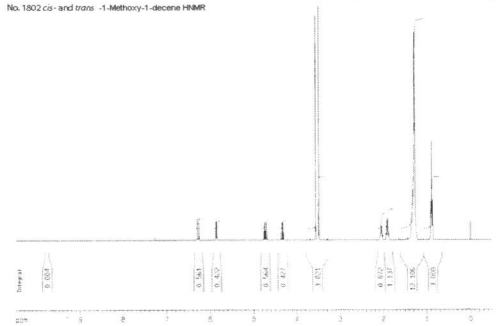

**1802   cis- and trans-1-Methoxy-1-decene (IR)**

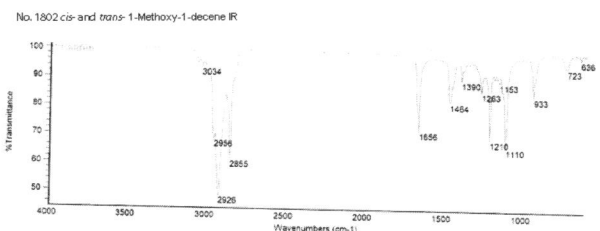

**1803 (E)-Tetradec-2-enal (MS)**

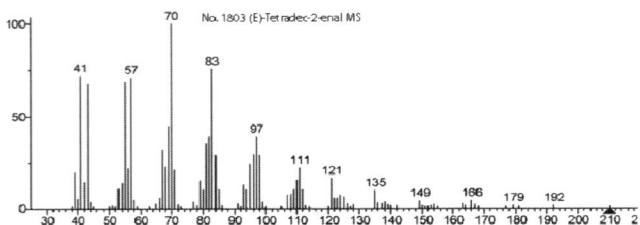

**1804 (E)-2-Pentenoic acid (1H-NMR)**

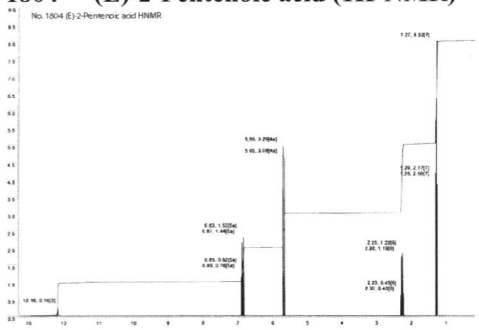

**1804 (E)-2-Pentenoic acid (MS)**

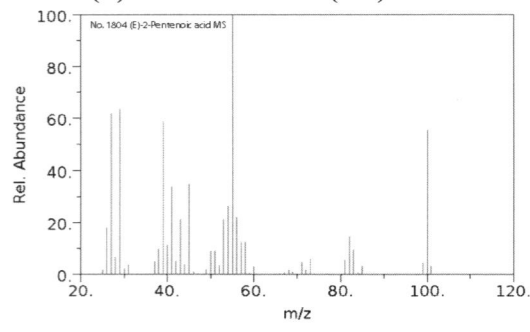

**1805 (E)-2-Octenoic acid (13C-NMR)**

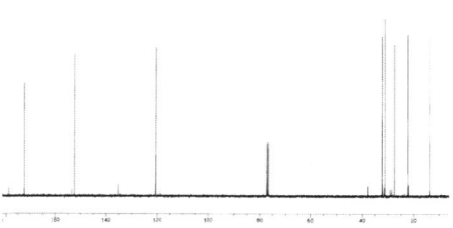

**1805 (E)-2-Octenoic acid (1H-NMR)**

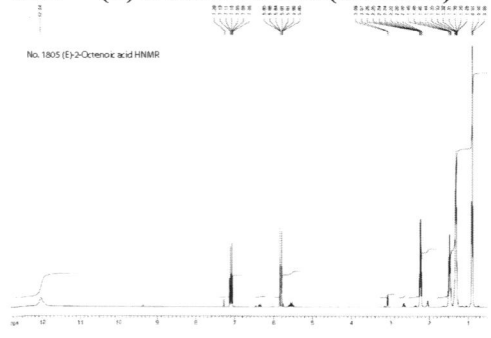

**1805 (E)-2-Octenoic acid (IR)**

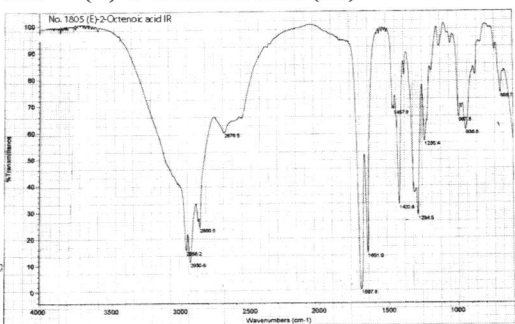

**1805 (E)-2-Octenoic acid (MS)**

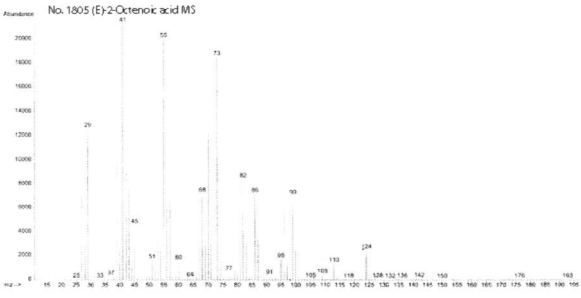

**1806 Ethyl trans-2-butenoate (MS)**

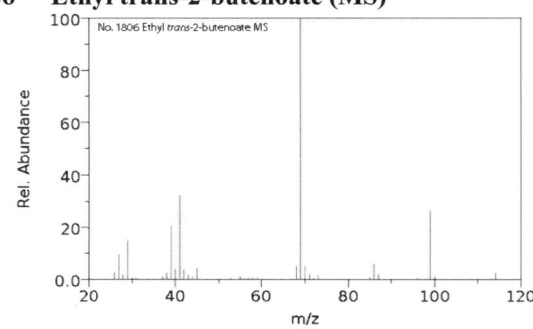

**1807 Hexyl 2-butenoate (NMR)**

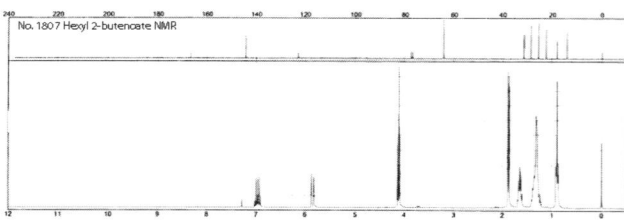

**1808 Ethyl trans-2-hexenoate (1H-NMR)**

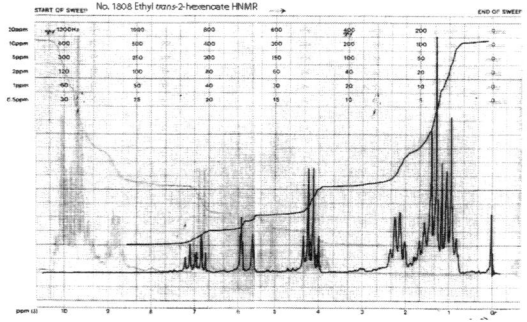

## 1808    Ethyl trans-2-hexenoate (IR)

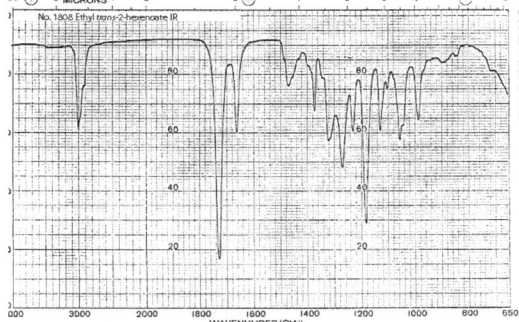

## 1809    (E,Z)-Methyl 2-hexenoate (1H-NMR)

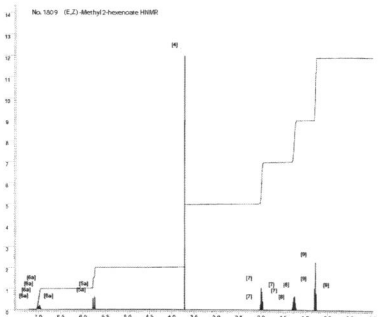

## 1810    Hexyl trans-2-hexenoate (1H-NMR)

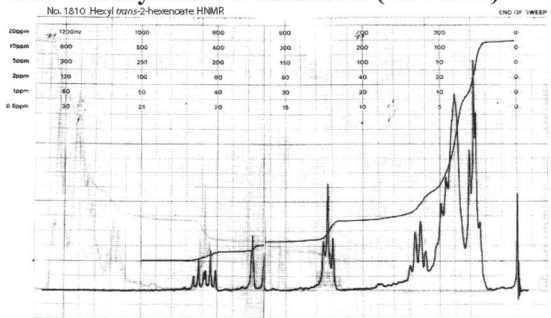

## 1810    Hexyl trans-2-hexenoate (IR)

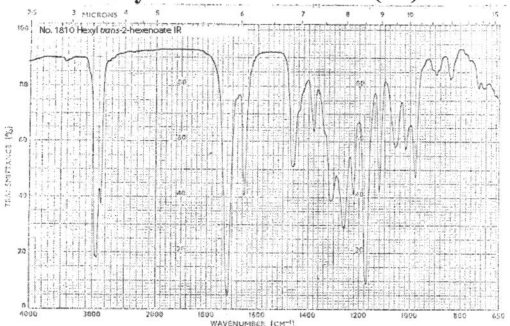

## 1811    Methyl trans-2-octenoate (1H-NMR)

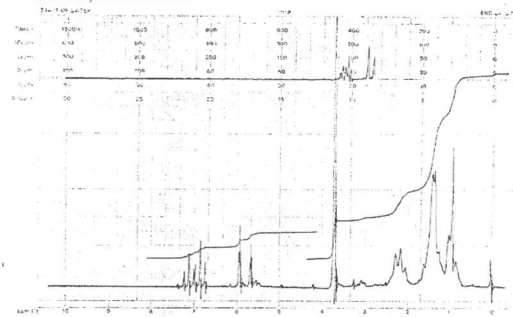

## 1811    Methyl trans-2-octenoate (IR)

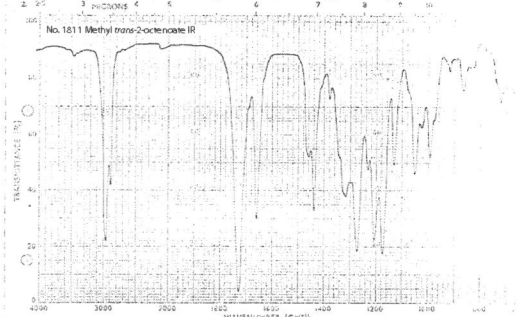

## 1811    Methyl trans-2-octenoate (MS)

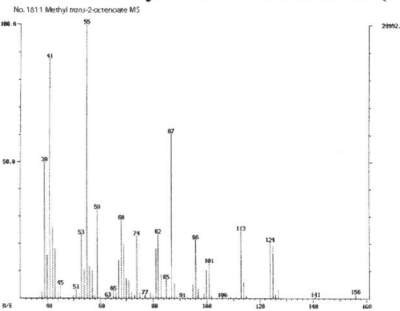

## 1812    Ethyl trans-2-octenoate (1H-NMR)

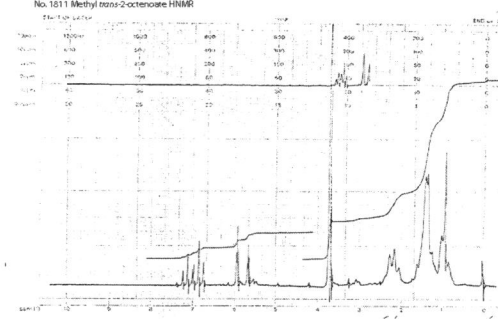

## 1812    Ethyl trans-2-octenoate (IR)

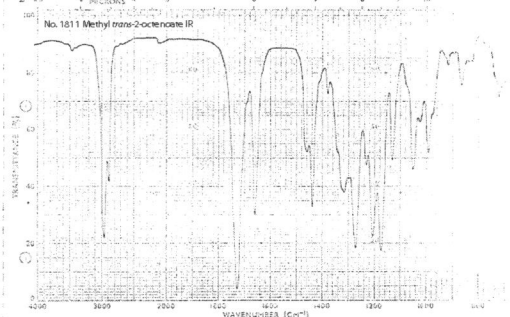

## 1812    Ethyl trans-2-octenoate (MS)

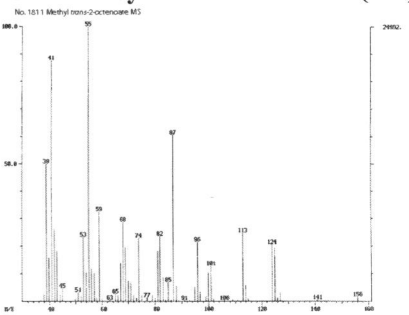

**1813 (E,Z)-Methyl 2-nonenoate (NMR)**

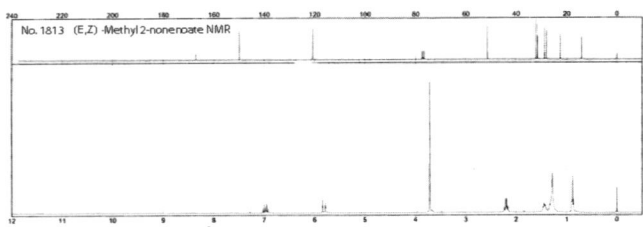

**1814 Ethyl trans-2-decenoate (1H-NMR)**

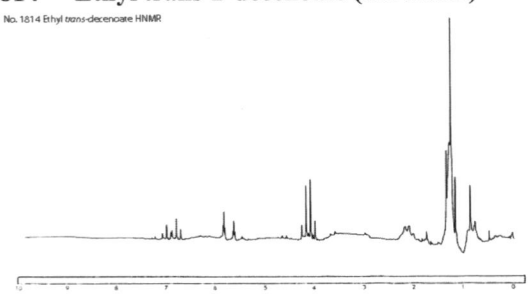

**1814 Ethyl trans-2-decenoate (IR)**

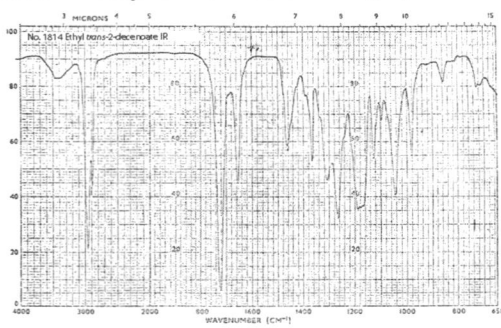

**1814 Ethyl trans-2-decenoate (MS)**

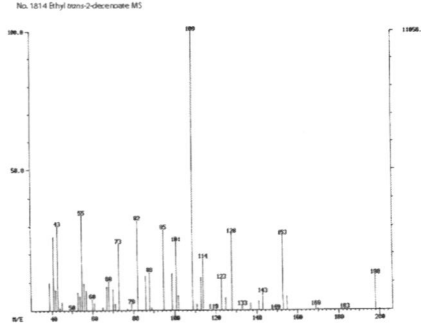

**1815 Ethyl (E)-2-methyl-2-pentenoate (1H-NMR)**

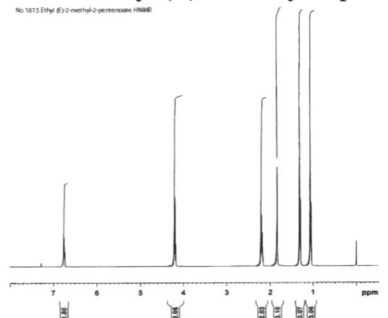

**1816 2-Methylbutyl 3-methyl-2-butenoate (1H-NMR)**

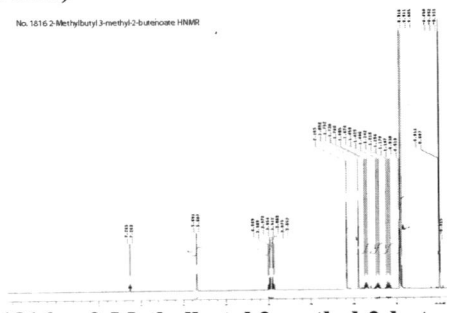

**1816 2-Methylbutyl 3-methyl-2-butenoate (MS)**

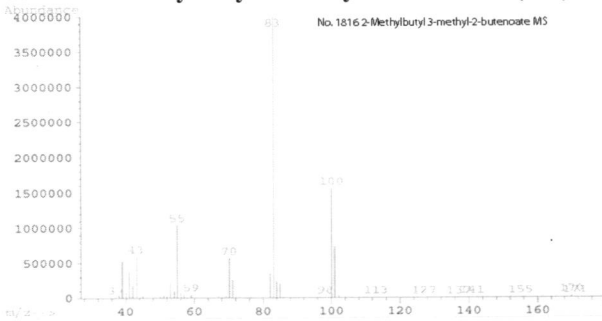

**1817 (+/-) (E,Z)-5-(2,2-Dimethylcyclopropyl)-3-methyl-2-pentenal (1H-NMR)**

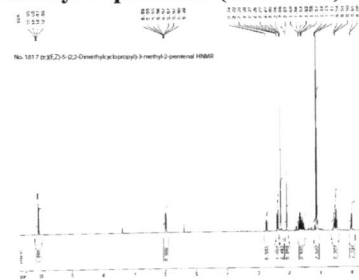

**1818 (E,Z)-4-Methylpent-2-enoic acid (MS)**

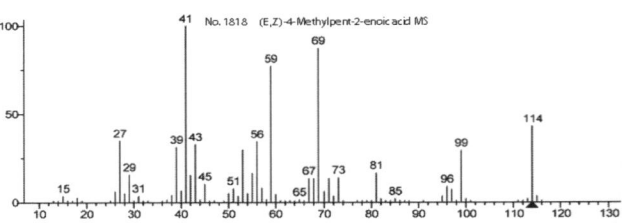

**1819 (+/-)-4-Ethyloctanal (1H-NMR)**

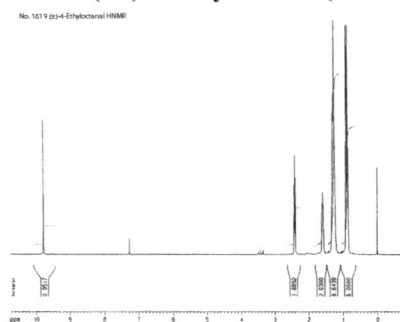

**1819 (+/-)-4-Ethyloctanal (IR)**

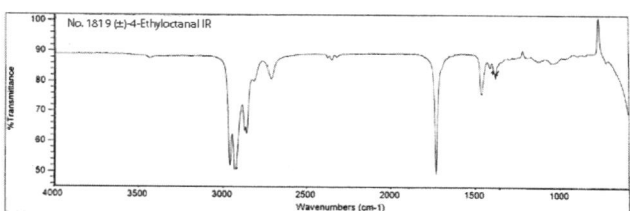

**1819 (+/-)-4-Ethyloctanal (MS)**

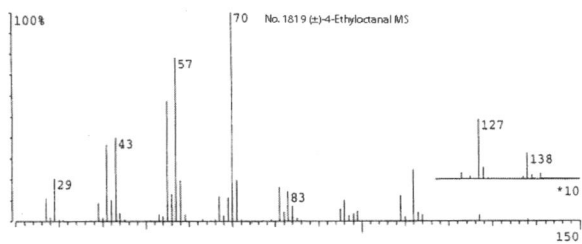

**1820 (E)-Geranyl 2-methylbutyrate (MS)**

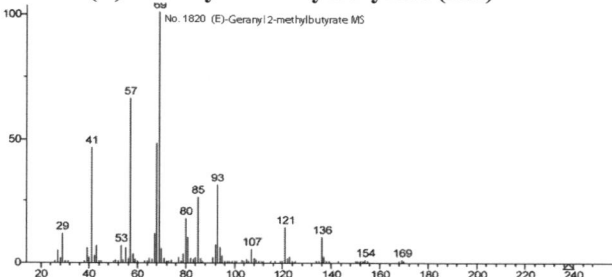

**1821 (E)-Geranyl valerate (MS)**

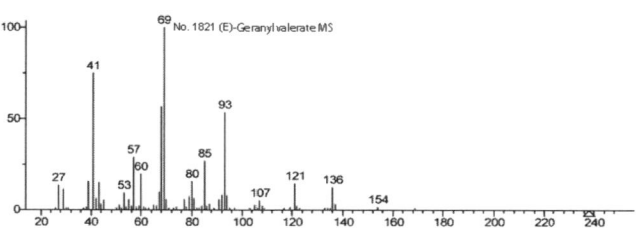

**1822 (E)-Geranyl tiglate (IR)**

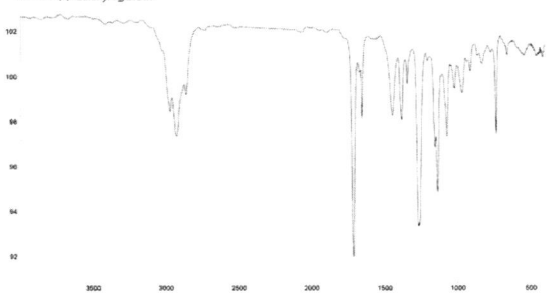

**1822 (E)-Geranyl tiglate (MS)**

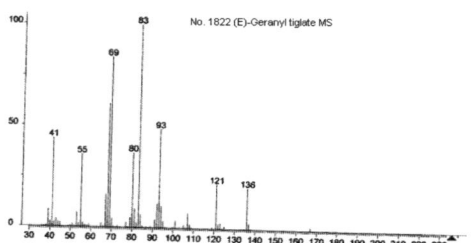

**1823 (E)-Citronellyl 2-methylbut-2-enoate (MS)**

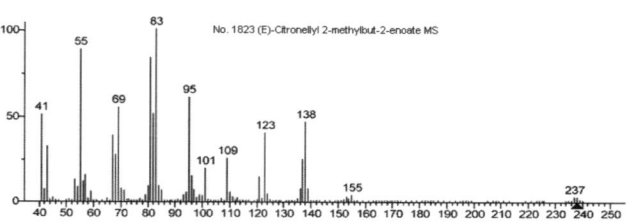

**1824 (E)-Ethyl tiglate (IR)**

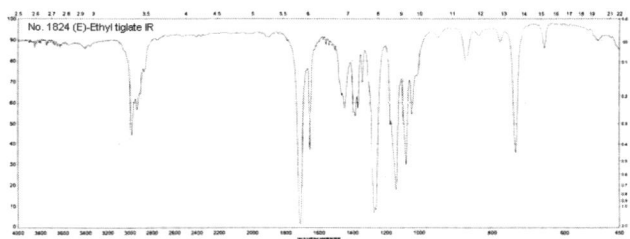

**1824 (E)-Ethyl tiglate (NMR)**

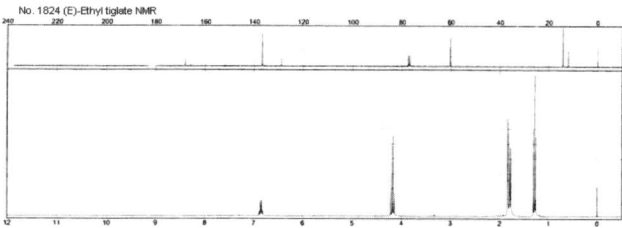

**1825 (E,Z)-Geranic acid (MS)**

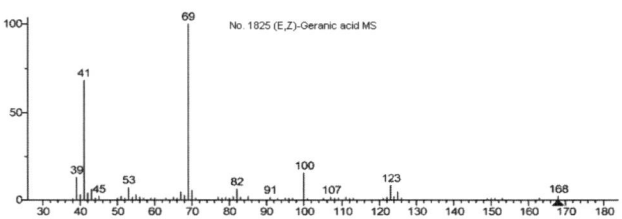

**1826 Prenyl formate (MS)**

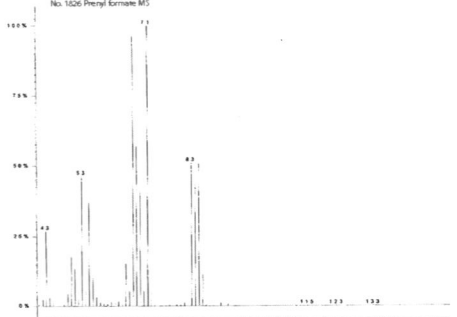

**1827 Prenyl acetate (MS)**

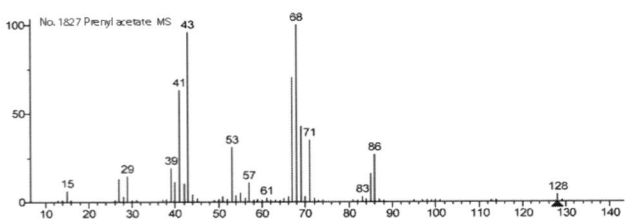

**1828 Prenyl isobutyrate (MS)**

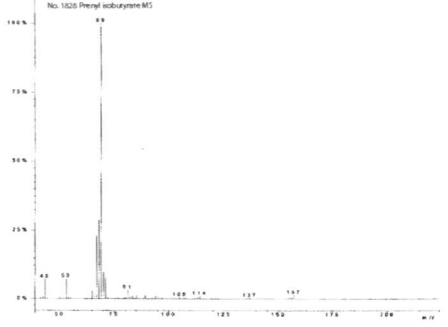

**1829 Prenyl caproate (MS)**

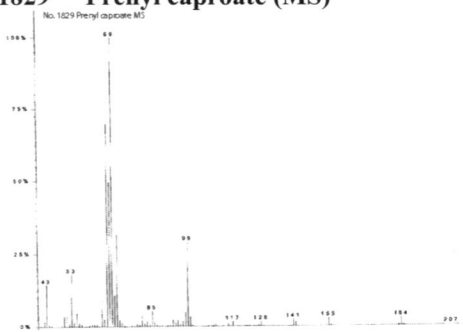

**1830 (+/-)-Dihydrofarnesol (1H-NMR)**

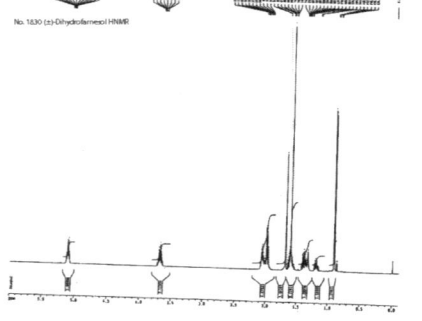

**1830 (+/-)-Dihydrofarnesol (IR)**

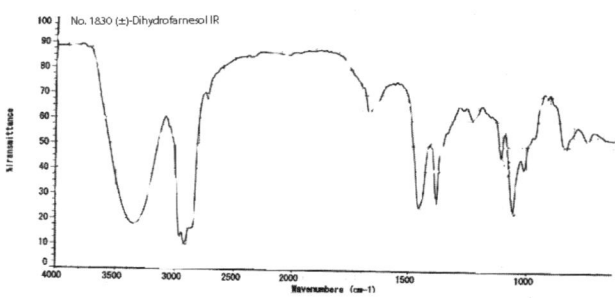

**1831 (E,Z)-3,7,11-Trimethyldodeca-2,6,10-trienyl acetate (MS)**

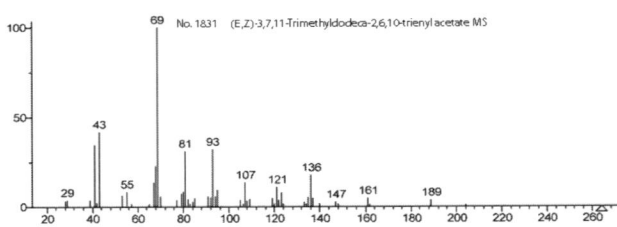

**1832 (E,Z)-Phytol (MS)**

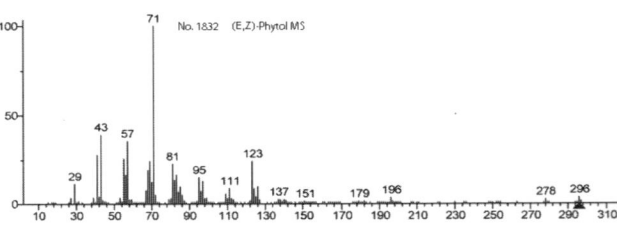

**1833 (E,Z)-Phytyl acetate (MS)**

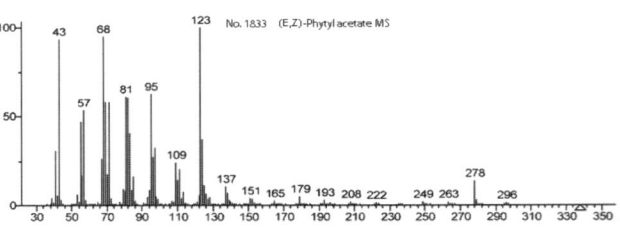

**1834 Methyl 2-methyl-2-propenoate (IR)**

**1835　Isopropenyl acetate (MS)**

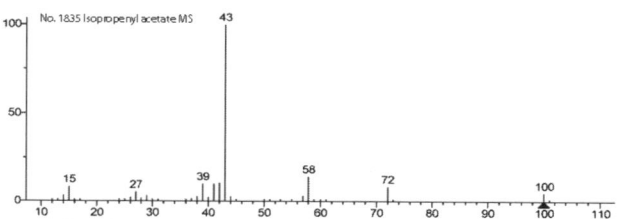

**1836　1-Octen-3-yl acetate (1H-NMR)**

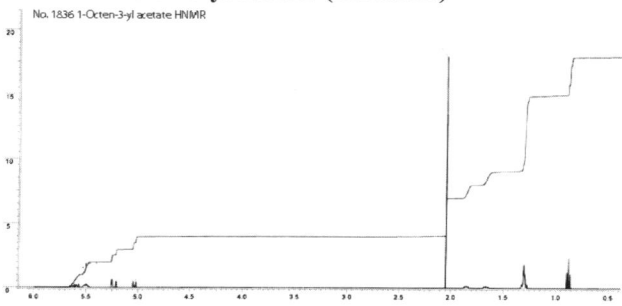

**1837　1-Octen-3-yl butyrate (1H-NMR)**

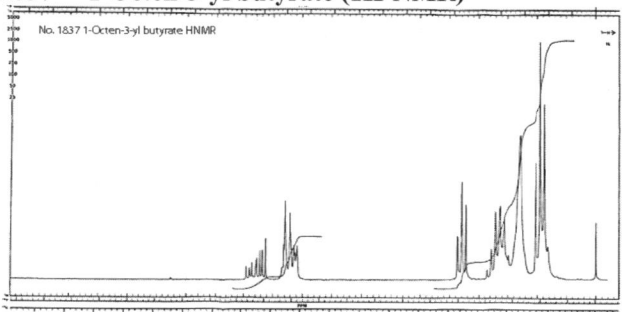

**1837　1-Octen-3-yl butyrate (IR)**

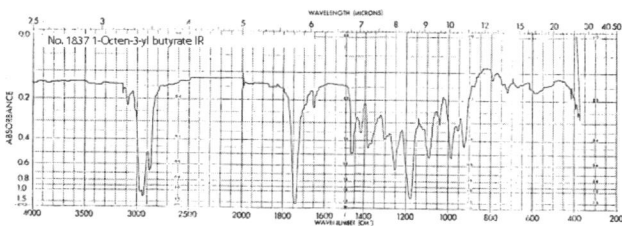

**1837　1-Octen-3-yl butyrate (MS)**

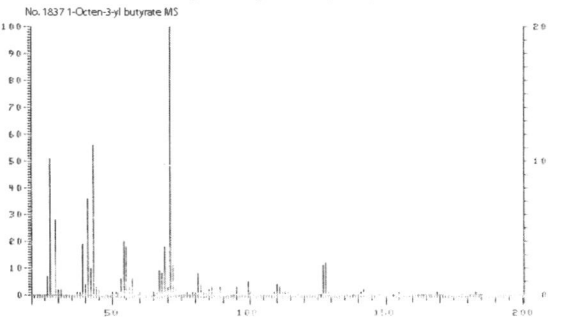

**1838　6-Methyl-5-hepten-2-yl acetate (1H-NMR)**

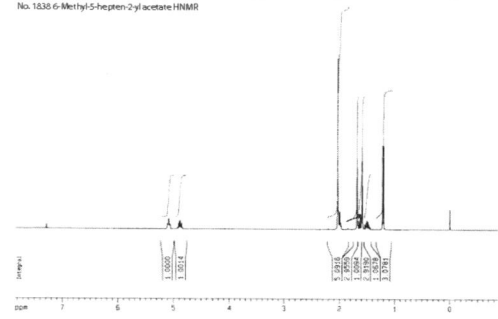

**1838　6-Methyl-5-hepten-2-yl acetate (IR)**

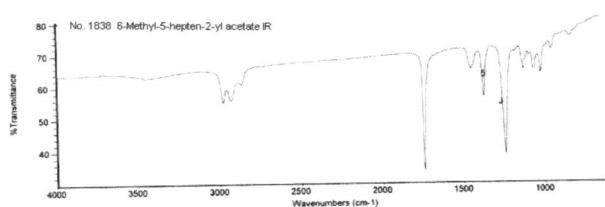

**1839　3-(Hydroxymethyl)-2-octanone (1H-NMR)**

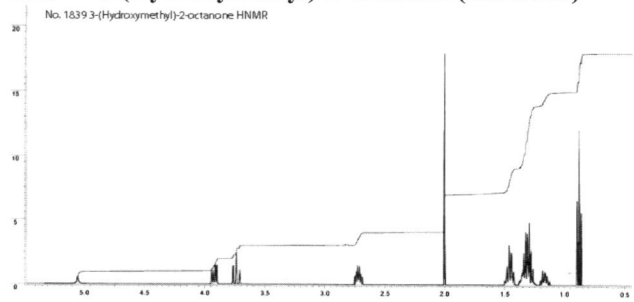

**1840　(+/-) [R-(E)]-5-Isopropyl-8-methylnona-6,8-dien-2-one (MS)**

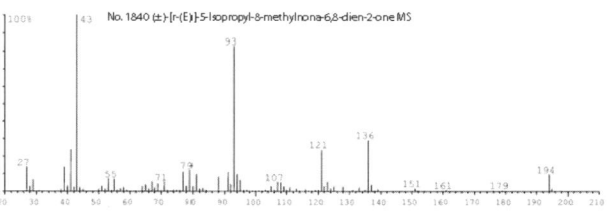

**1841　(+/-)-cis- and trans-4,8-Dimethyl-3,7-nonadien-2-ol (1H-NMR)**

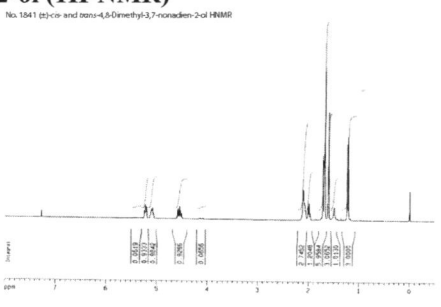

**1841 (+/-)-cis- and trans-4,8-Dimethyl-3,7-nonadien-2-ol (IR)**

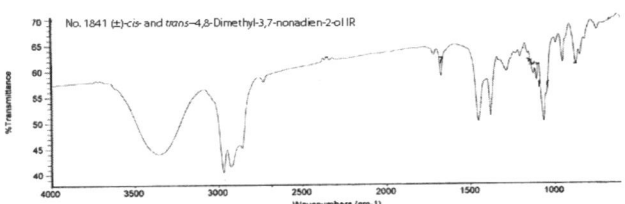

**1842 (+/-)-1-Hepten-3-ol (13C-NMR)**

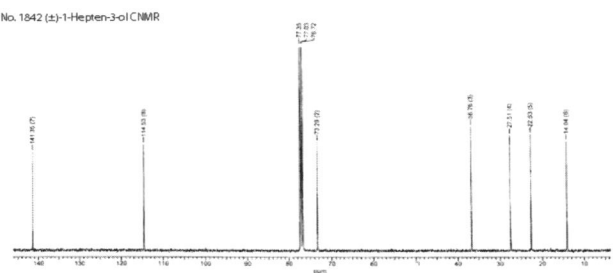

**1842 (+/-)-1-Hepten-3-ol (1H-NMR)**

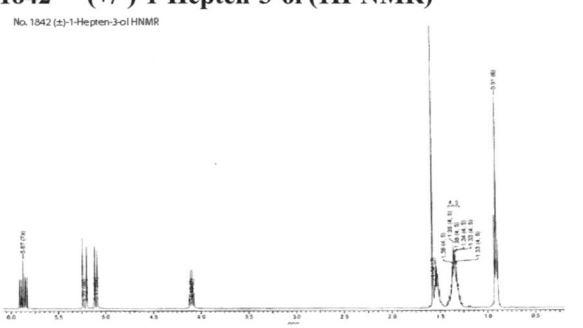

**1842 (+/-)-1-Hepten-3-ol (IR)**

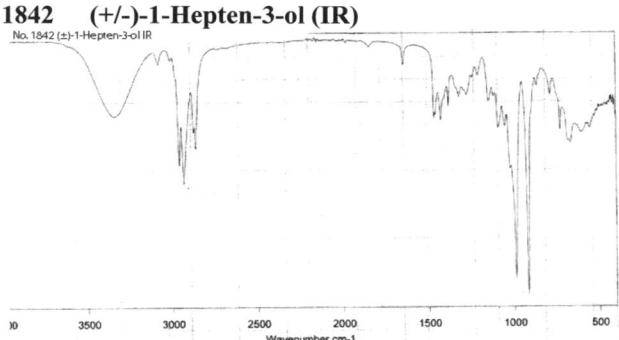

**1842 (+/-)-1-Hepten-3-ol (MS)**

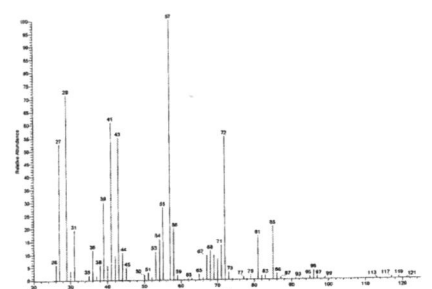

**1843 (E,Z)-4-Octen-3-one (1H-NMR)**

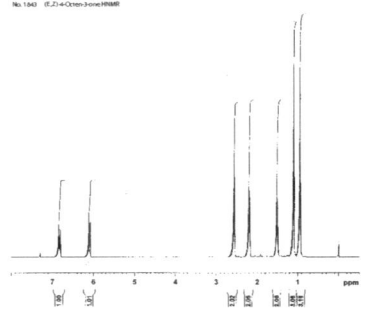

**1844 (E)-2-Nonen-4-one (1H-NMR)**

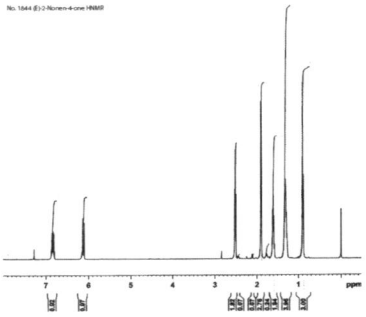

**1845 (E)-5-Nonen-2-one (1H-NMR)**

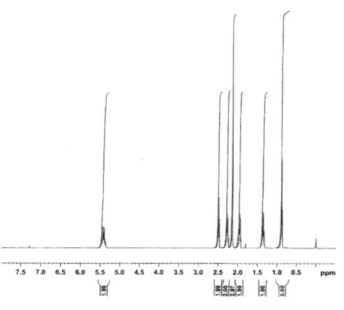

**1846 (Z)-3-Hexenyl 2-oxopropionate (1H-NMR)**

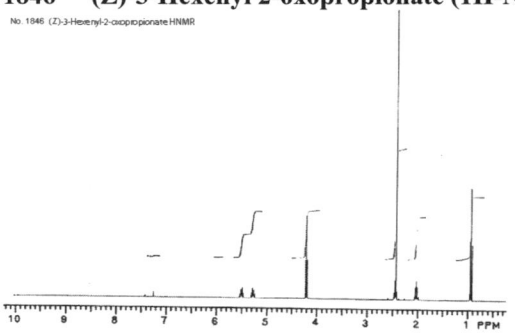

**1846 (Z)-3-Hexenyl 2-oxopropionate (IR)**

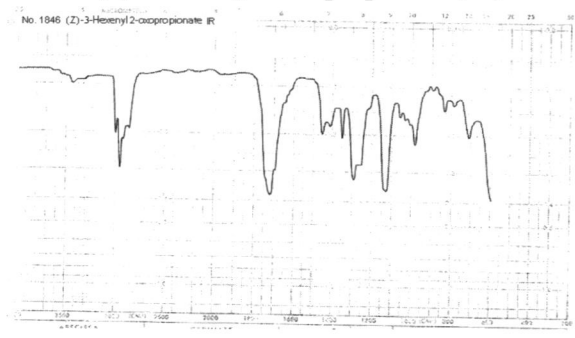

**1847 (+/-)-cis and trans-4,8-Dimethyl-3,7-nonadien-2-yl acetate (1H-NMR)**

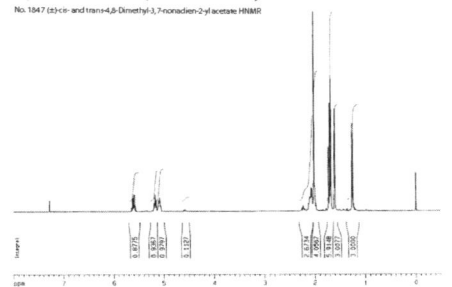

**1847 (+/-)-cis and trans-4,8-Dimethyl-3,7-nonadien-2-yl acetate (IR)**

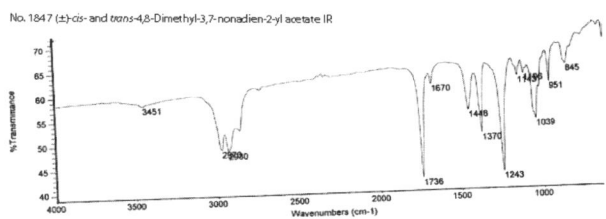

**1848 (E)-1,5-Octadien-3-one (MS)**

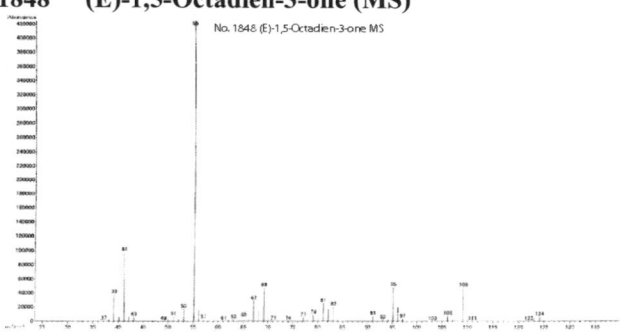

**1849 10-Undecen-2-one (MS)**

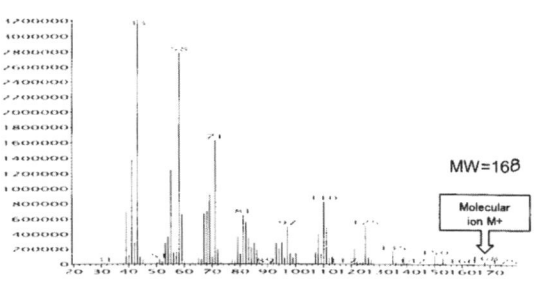

**1850 2,4-Dimethyl-4-nonanol (MS)**

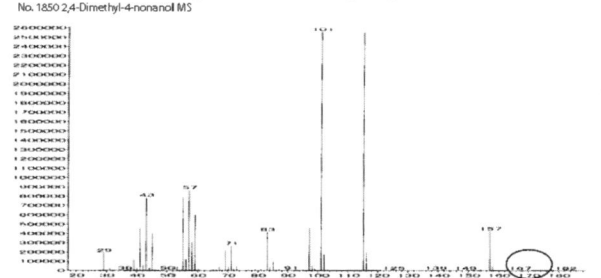

**1851 8-Nonen-2-one (MS)**

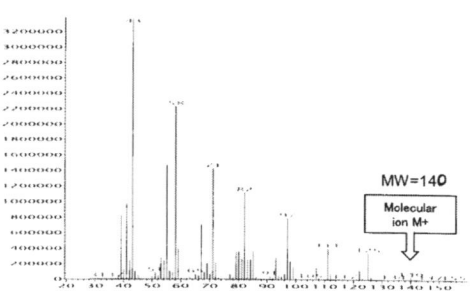

**1852 Menthyl valerate (1H-NMR)**

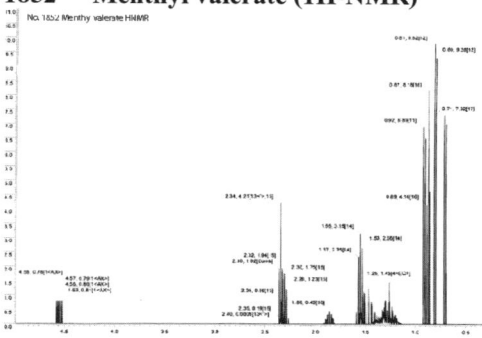

**1852 Menthyl valerate (MS)**

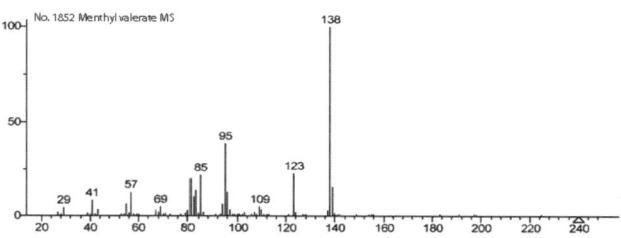

**1853 2-(l-Menthoxy)ethanol (1H-NMR)**

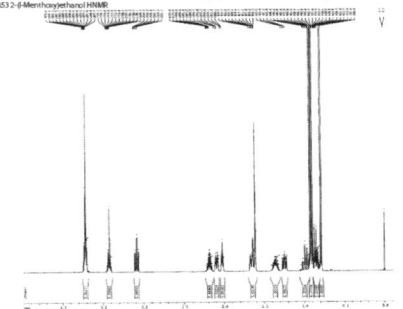

**1854 l-Menthyl acetoacetate (1H-NMR)**

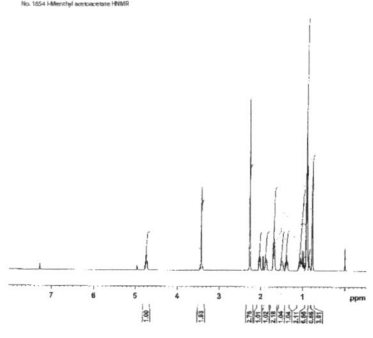

**1855  l-Menthyl (R,S)-3-hydroxybutyrate (1H-NMR)**

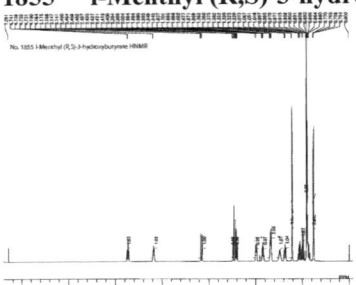

**1855  l-Menthyl (R,S)-3-hydroxybutyrate (IR)**

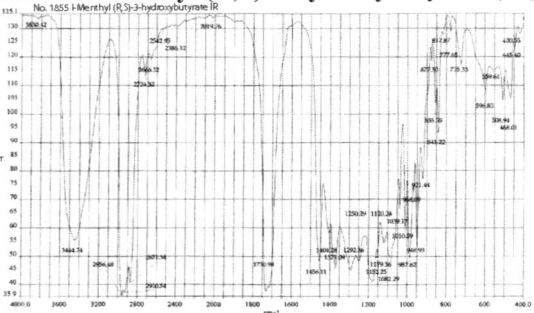

**1856  l-Piperitone (1H-NMR)**

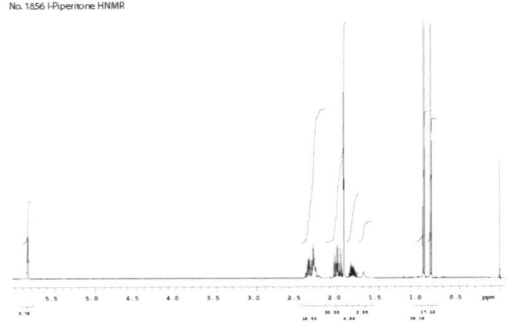

**1856  l-Piperitone (IR)**

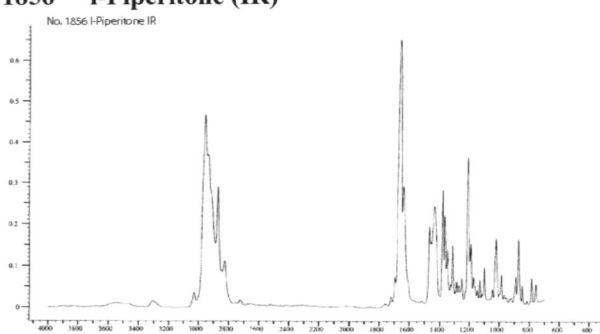

**1856  l-Piperitone (MS)**

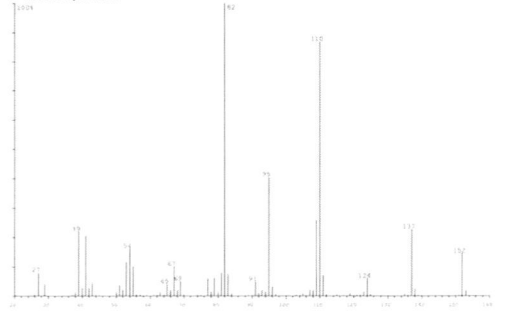

**1857  2,6,6-Trimethylcyclohex-2-ene-1,4-dione (IR)**

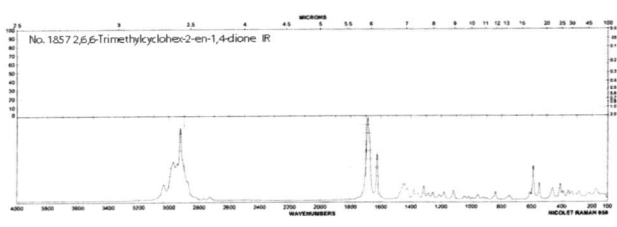

**1857  2,6,6-Trimethylcyclohex-2-ene-1,4-dione (NMR)**

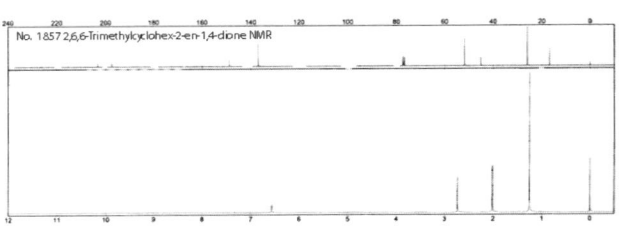

**1858  Menthyl pyrrolidone carboxylate (13C-NMR)**

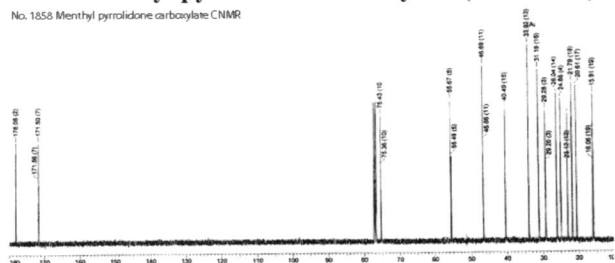

**1858  Menthyl pyrrolidone carboxylate (IR)**

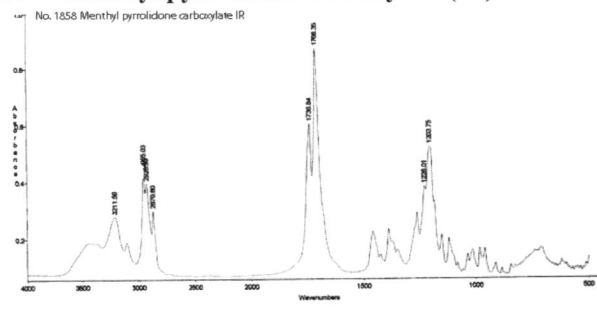

**1858  Menthyl pyrrolidone carboxylate (MS)**

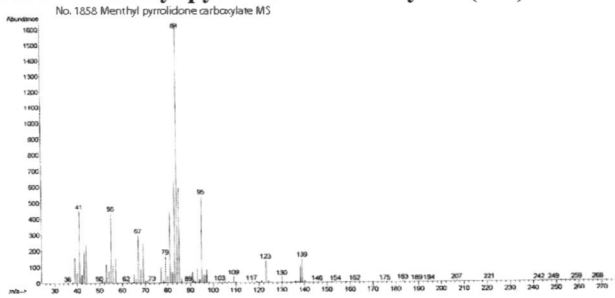

**1859  3,9-Dimethyl-6-(1-methylethyl)-1,4-dioxaspiro[4.5]decan-2-one (13C-NMR)**

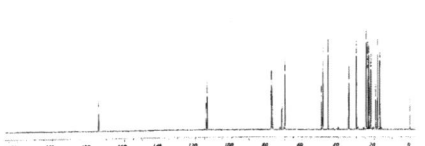

**1859  3,9-Dimethyl-6-(1-methylethyl)-1,4-dioxaspiro[4.5]decan-2-one (1H-NMR)**

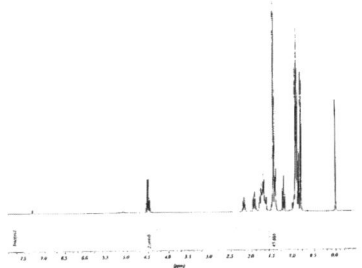

**1859  3,9-Dimethyl-6-(1-methylethyl)-1,4-dioxaspiro[4.5]decan-2-one (IR)**

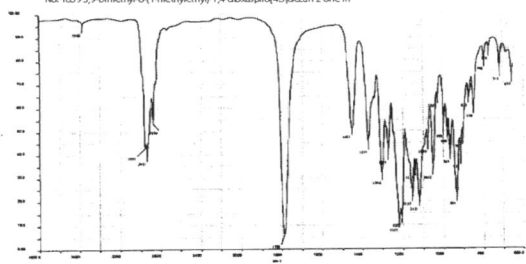

**1859  3,9-Dimethyl-6-(1-methylethyl)-1,4-dioxaspiro[4.5]decan-2-one (MS)**

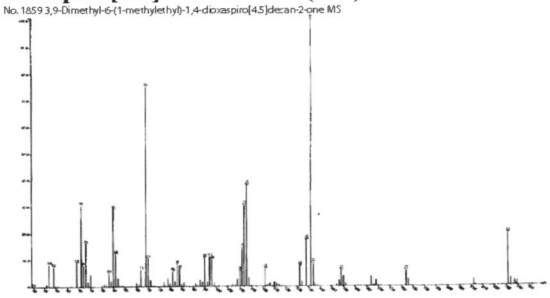

**1860  8-p-Menthene-1,2-diol (MS)**

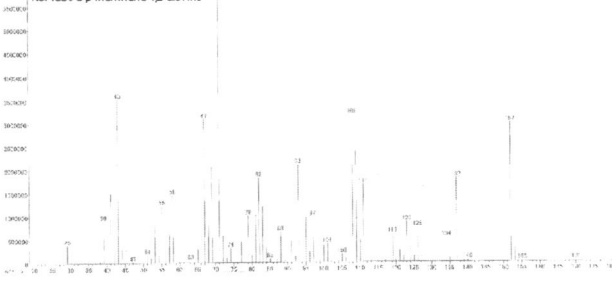

**1861  d-2,8-p-Menthadien-1-ol (MS)**

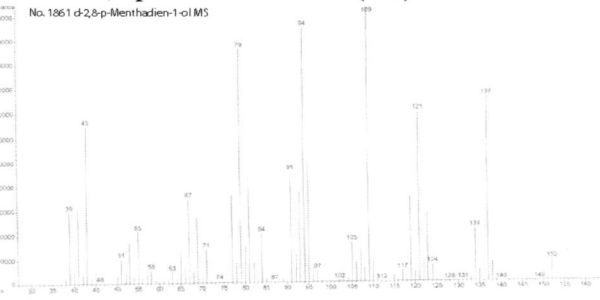

**1862  Dehydronootkatone (1H-NMR)**

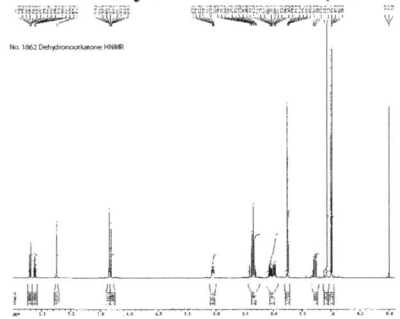

**1863  Isobornyl isobutyrate (MS)**

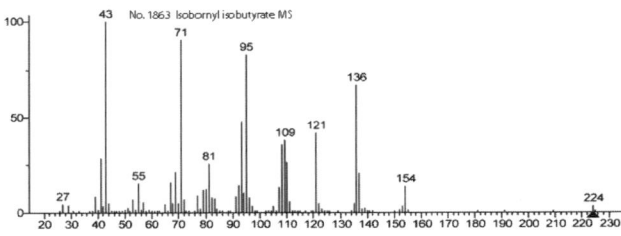

**1864  l-Bornyl acetate (1H-NMR)**

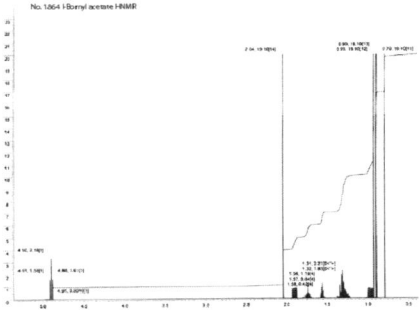

**1865  Thujyl alcohol (1H-NMR)**

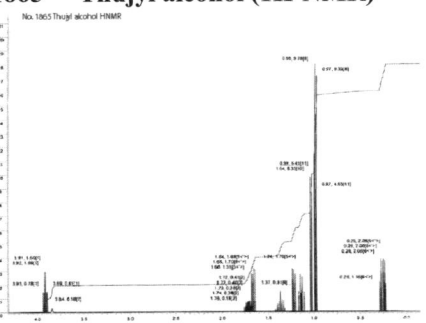

**1866    Vetiverol (1H-NMR)**

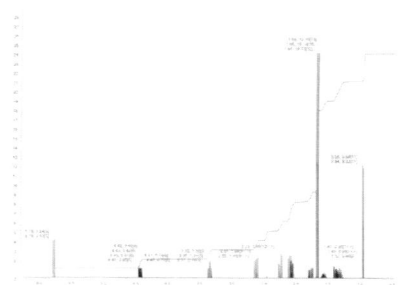

**1867    Vetiveryl acetate (1H-NMR)**

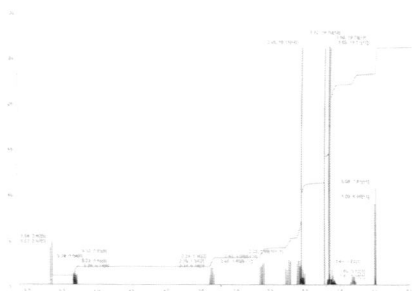

**1868    3-Pinanone (1H-NMR)**

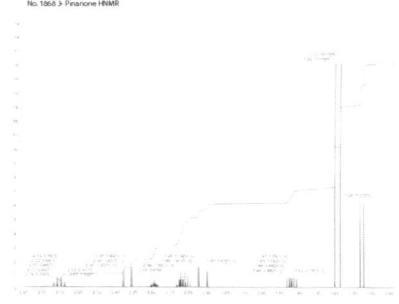

**1868    3-Pinanone (MS)**

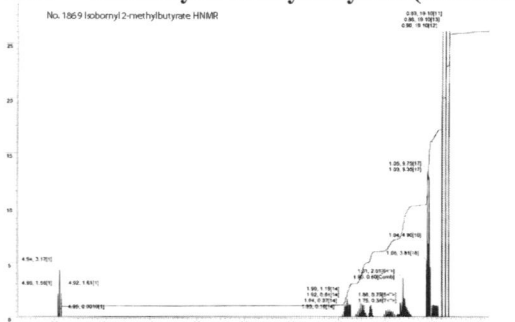

**1869    Isobornyl 2-methylbutyrate (1H-NMR)**

**1869    Isobornyl 2-methylbutyrate (MS)**

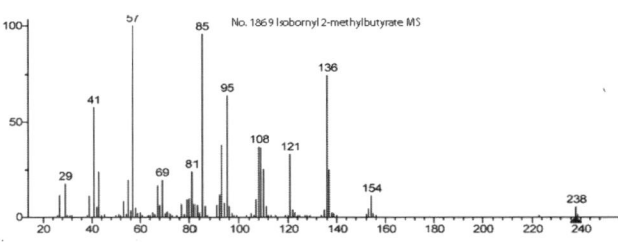

**1870    Verbenone (1H-NMR)**

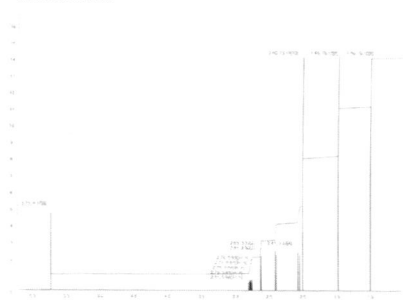

**1870    Verbenone (MS)**

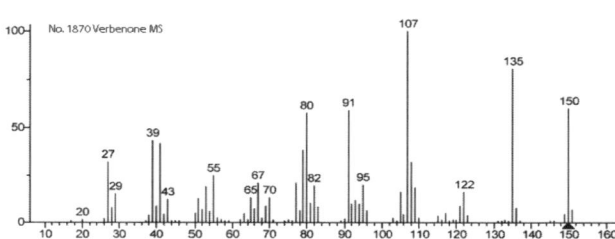

**1871    Methyl hexanoate (NMR)**

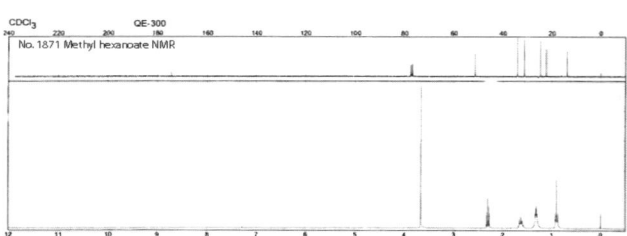

**1872    Hexyl heptanoate (MS)**

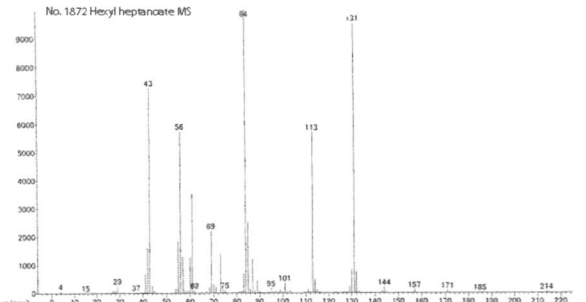

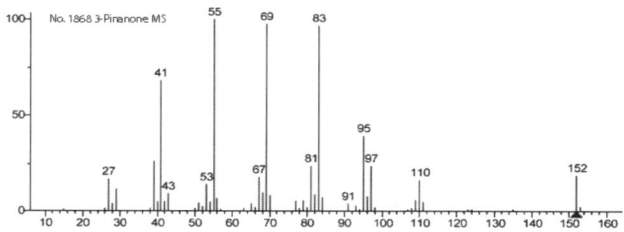

**1873    Hexyl nonanoate (MS)**

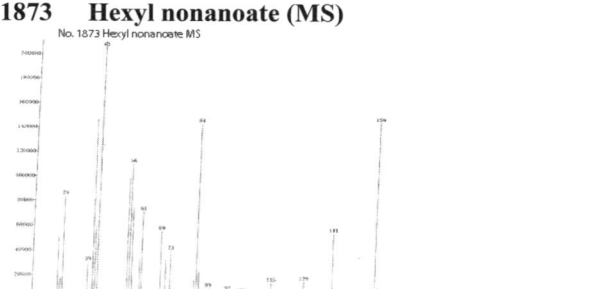

**1874    Hexyl decanoate (MS)**

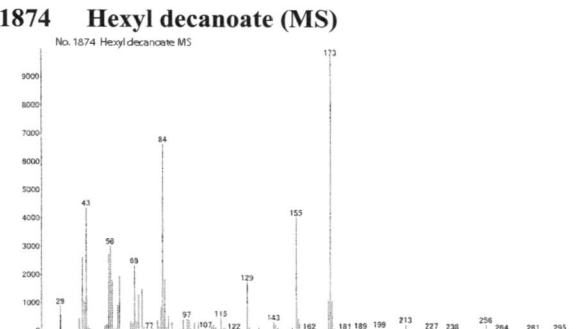

**1875    Heptyl heptanoate (MS)**

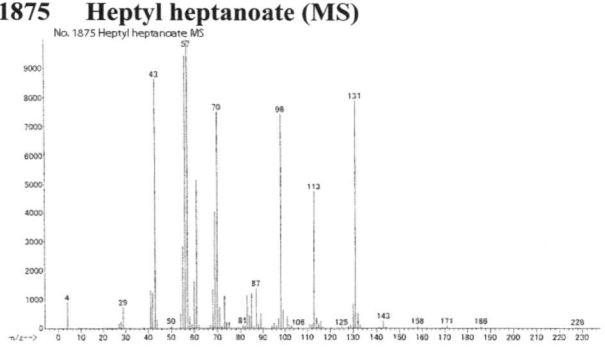

**1876    Dodecyl propionate (1H-NMR)**

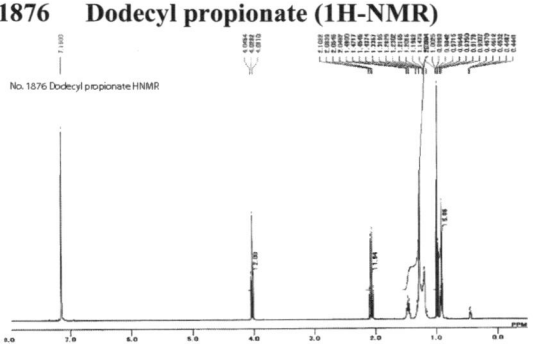

**1876    Dodecyl propionate (IR)**

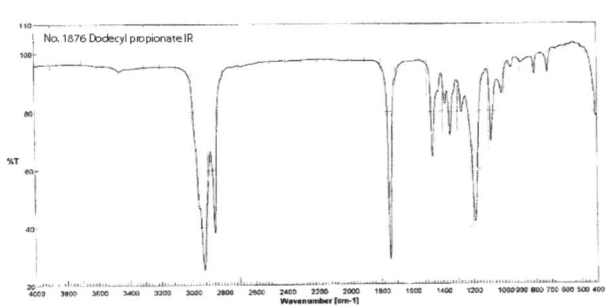

**1876    Dodecyl propionate (MS)**

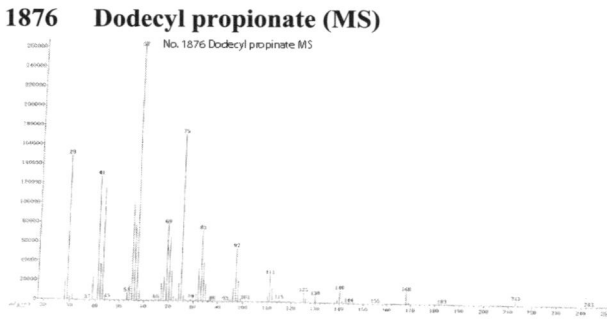

**1877    Dodecyl butyrate (13C-NMR)**

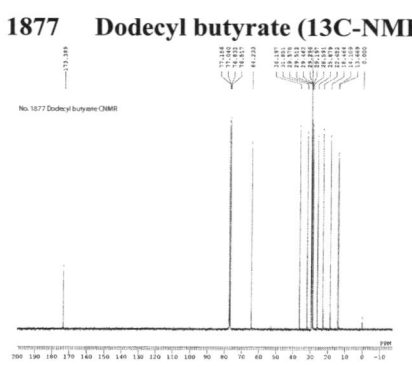

**1877    Dodecyl butyrate (1H-NMR)**

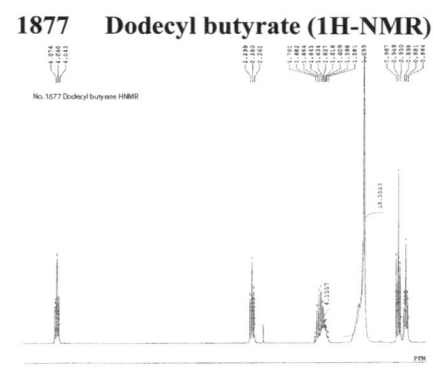

**1877    Dodecyl butyrate (MS)**

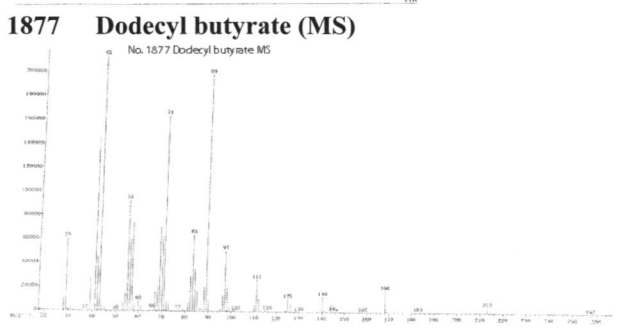

**1878    4-Hydroxy-3,5-dimethoxy benzaldehyde (IR)**

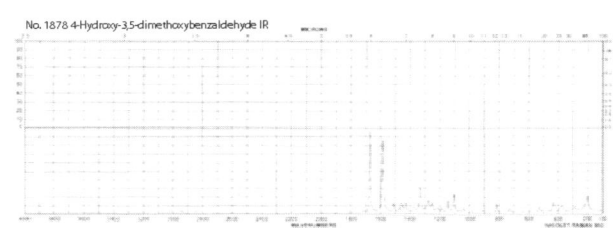

**1878   4-Hydroxy-3,5-dimethoxy benzaldehyde (NMR)**

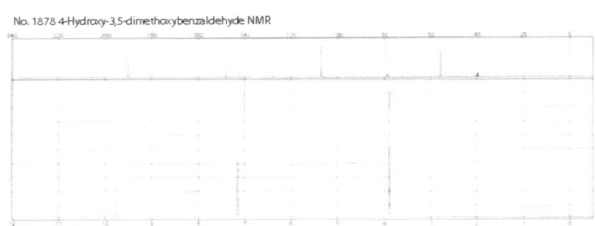

**1879   Vanillin 3-(l-menthoxy)propane-1,2-diol acetal (1H-NMR)**

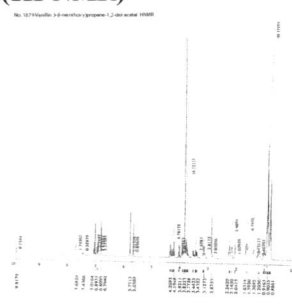

**1879   Vanillin 3-(l-menthoxy)propane-1,2-diol acetal (MS)**

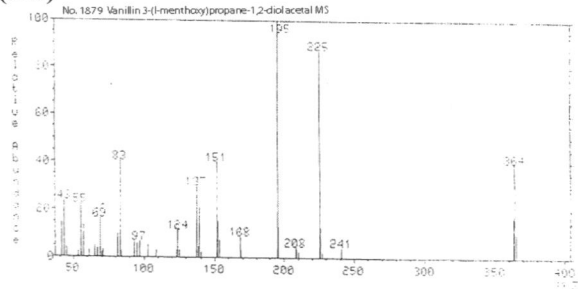

**1880   Sodium 4-methoxybenzoyloxyacetate (1H-NMR)**

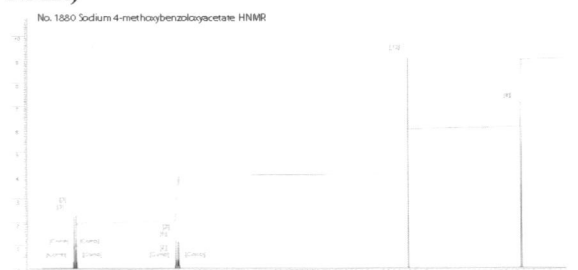

**1881   Divanillin (1H-NMR)**

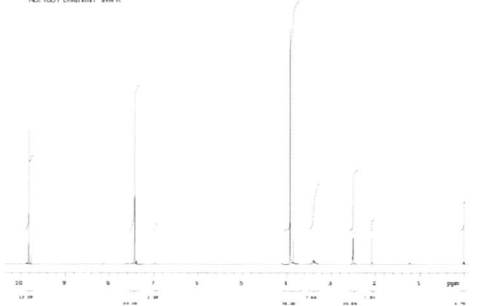

**1882   Vanillin propylene glycol acetal (1H-NMR)**

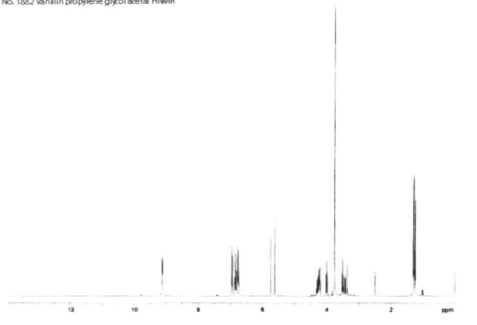

**1883   4-Methoxybenzoyloxyacetic acid (1H-NMR)**

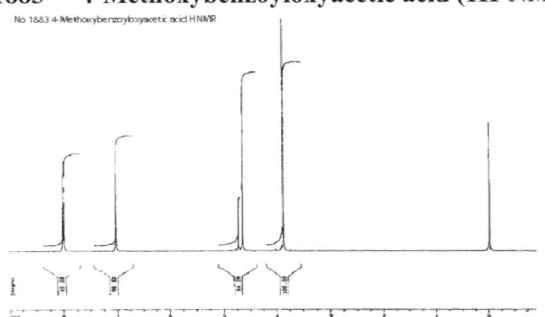

**1883   4-Methoxybenzoyloxyacetic acid (IR)**

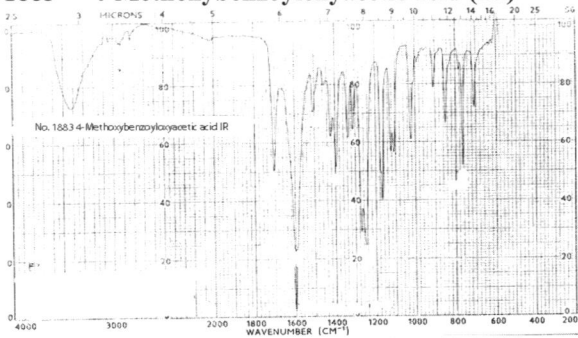

**1883   4-Methoxybenzoyloxyacetic acid (MS)**

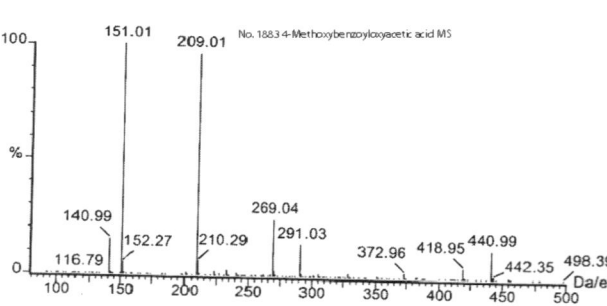

**1884   Methyl isothiocyanate (MS)**

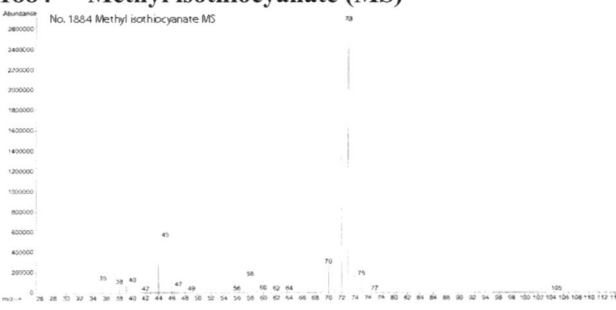

**1885 Ethyl isothiocyanate (MS)**

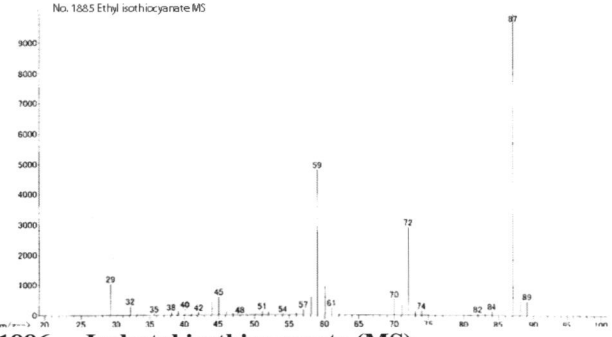

**1886 Isobutyl isothiocyanate (MS)**

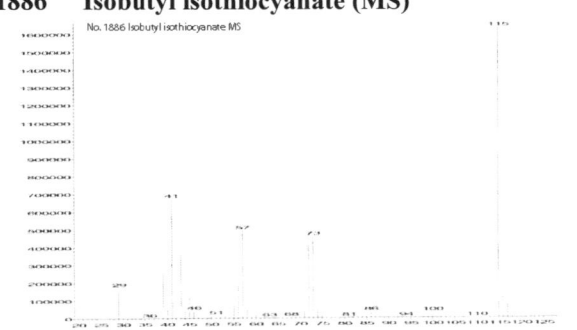

**1887 Isoamyl isothiocyanate (MS)**

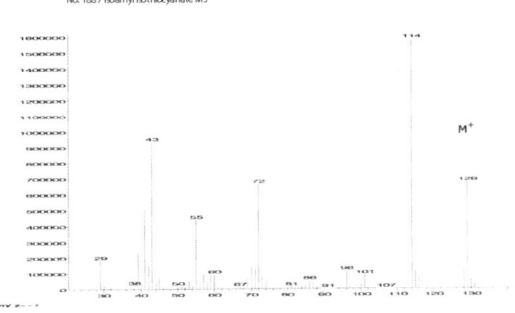

**1888 Isopropyl isothiocyanate (MS)**

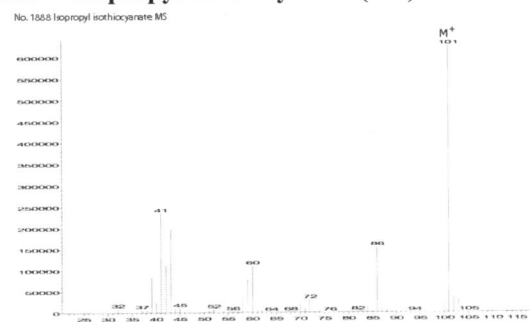

**1889 3-Butenyl isothiocyanate (MS)**

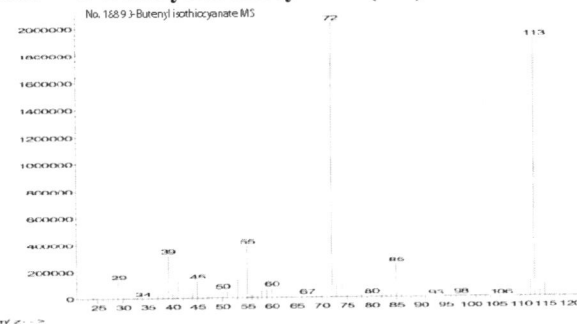

**1890 2-Butyl isothiocyanate (MS)**

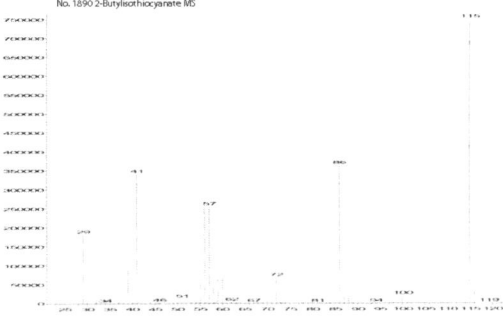

**1891 Amyl isothiocyanate (MS)**

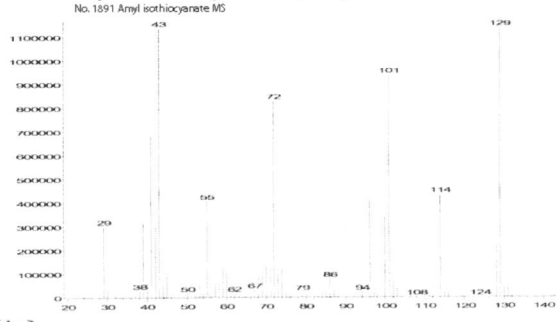

**1892 4-(Methylthio)butyl isothiocyanate (MS)**

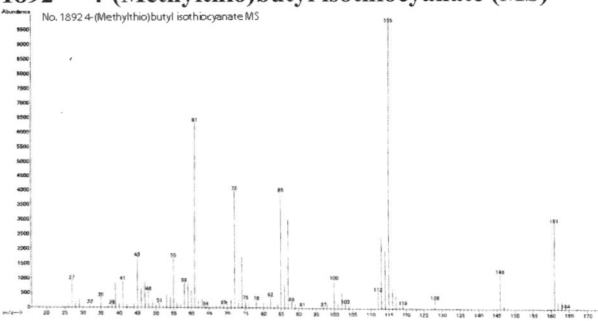

**1893 4-Pentenyl isothiocyanate (MS)**

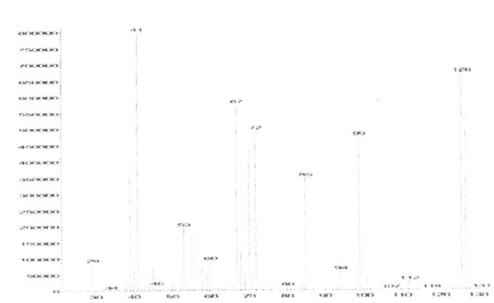

**1894 5-Hexenyl isothiocyanate (MS)**

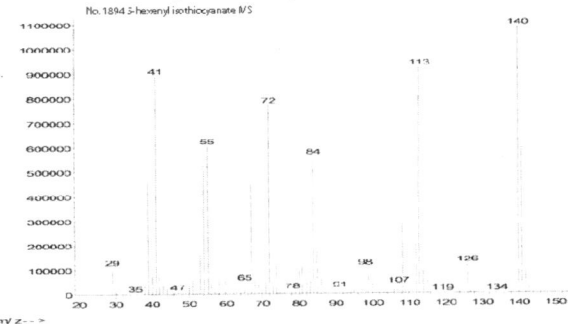

**1895  Hexyl isothiocyanate (MS)**

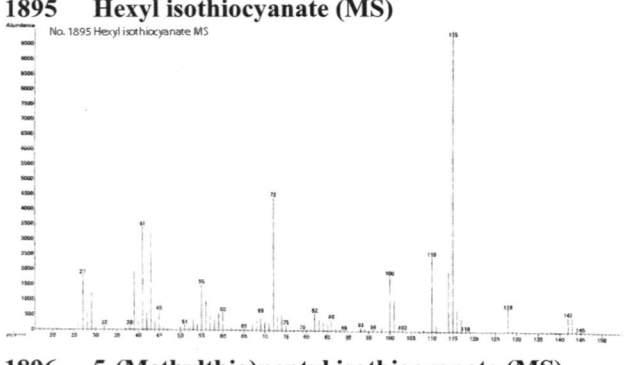

**1897  6-(Methylthio)hexyl isothiocyanate (MS)**

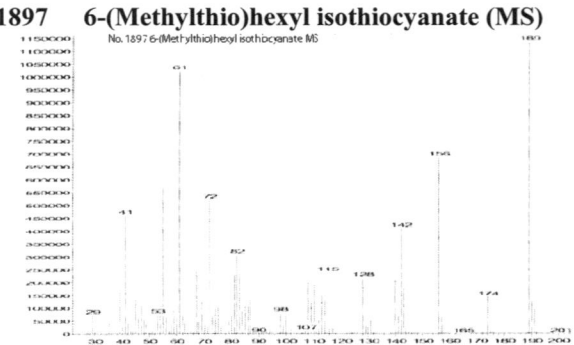

**1896  5-(Methylthio)pentyl isothiocyanate (MS)**

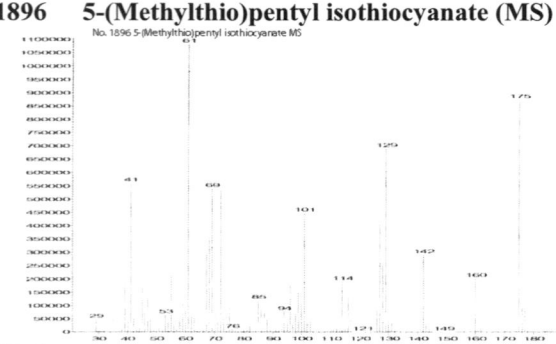

# Index: Specifications of certain flavourings

| Name | Page |
|---|---|
| Apiole | 91 |
| Amyl isothiocyanate | 106 |
| l-Bornyl acetate | 102 |
| 3-Butenyl isothiocyanate | 106 |
| 2-Butyl isothiocyanate | 106 |
| (E)-Citronellyl 2-methylbut-2-enoate | 96 |
| (E)-2-Decen-1-ol | 92 |
| Dehydronootkatone | 102 |
| (+/-)-Dihydrofarnesol | 97 |
| (+/-) (E,Z)-5-(2,2-Dimethylcyclopropyl)-3-methyl-2-pentenal | 95 |
| 3,9-Dimethyl-6-(1-methylethyl)-1,4-dioxaspiro[4.5]decan-2-one | 102 |
| (+/-)-cis- and trans-4,8-Dimethyl-3,7-nonadien-2-ol | 99 |
| (+/-)-cis and trans-4,8-Dimethyl-3,7-nonadien-2-yl acetate | 100 |
| 2,4-Dimethyl-4-nonanol | 100 |
| Divanillin | 105 |
| Dodecyl butyrate | 104 |
| Dodecyl propionate | 104 |
| Elemicin | 91 |
| Estragole | 91 |
| Ethyl trans-2-butenoate | 94 |
| Ethyl trans-2-decenoate | 95 |
| Ethyl trans-2-hexenoate | 94 |
| Ethyl isothiocyanate | 106 |
| Ethyl (E)-2-methyl-2-pentenoate | 95 |
| Ethyl trans-2-octenoate | 95 |
| (E)-Ethyl tiglate | 96 |
| (+/-)-4-Ethyloctanal | 96 |
| (E,Z)-Geranic acid | 97 |
| (E)-Geranyl 2-methylbutyrate | 96 |
| (E)-Geranyl tiglate | 96 |
| (E)-Geranyl valerate | 96 |
| (+/-)-1-Hepten-3-ol | 99 |
| Hept-trans-2-en-1-yl acetate | 92 |
| (E,Z)-Hept-2-en-1-yl isovalerate | 92 |
| Heptyl heptanoate | 104 |
| trans-2-Hexenal glyceryl acetal | 93 |
| trans-2-Hexenal propylene glycol acetal | 93 |
| 5-Hexenyl isothiocyanate | 107 |
| trans-2-Hexenyl 2-methylbutyrate | 92 |
| (E)-2-Hexenyl octanoate | 92 |
| (Z)-3-Hexenyl 2-oxopropionate | 100 |
| Hexyl 2-butenoate | 94 |
| Hexyl decanoate | 104 |
| Hexyl heptanoate | 104 |
| Hexyl trans-2-hexenoate | 94 |
| Hexyl isothiocyanate | 107 |
| Hexyl nonanoate | 104 |
| 4-Hydroxy-3,5-dimethoxy benzaldehyde | 105 |
| 3-(Hydroxymethyl)-2-octanone | 99 |
| Isoamyl isothiocyanate | 106 |
| Isobornyl isobutyrate | 102 |
| Isobornyl 2-methylbutyrate | 103 |
| Isobutyl isothiocyanate | 106 |
| Isopropenyl acetate | 98 |
| Isopropyl isothiocyanate | 106 |
| (+/-) [R-(E)]-5-Isopropyl-8-methylnona-6,8-dien-2-one | 99 |
| d-2,8-p-Menthadien-1-ol | 102 |
| 8-p-Menthene-1,2-diol | 101 |
| 2-(l-Menthoxy)ethanol | 101 |
| l-Menthyl acetoacetate | 101 |
| Menthyl pyrrolidone carboxylate | 102 |
| l-Menthyl (R,S)-3-hydroxybutyrate | 101 |
| Menthyl valerate | 101 |
| 4-Methoxybenzoyloxyacetic acid | 105 |
| cis- and trans-1-Methoxy-1-decene | 93 |
| Methyl eugenol | 91 |
| Methyl hexanoate | 104 |
| (E,Z)-Methyl 2-hexenoate | 94 |
| Methyl isothiocyanate | 105 |
| Methyl 2-methyl-2-propenoate | 98 |
| (E,Z)-Methyl 2-nonenoate | 95 |
| Methyl trans-2-octenoate | 94 |
| 2-Methylbutyl 3-methyl-2-butenoate | 95 |
| 6-Methyl-5-hepten-2-yl acetate | 99 |
| (E,Z)-4-Methylpent-2-enoic acid | 96 |
| 4-(Methylthio)butyl isothiocyanate | 107 |
| 6-(Methylthio)hexyl isothiocyanate | 107 |
| 5-(Methylthio)pentyl isothiocyanate | 107 |
| Myristicin | 91 |
| (E)-2-Nonen-4-one | 100 |
| (E)-5-Nonen-2-one | 100 |
| 8-Nonen-2-one | 101 |
| (E)-1,5-Octadien-3-one | 100 |
| (E)-2-Octenoic acid | 94 |
| (E,Z)-4-Octen-3-one | 99 |
| 1-Octen-3-yl acetate | 98 |
| 1-Octen-3-yl butyrate | 98 |
| (E)-2-Pentenoic acid | 93 |
| (Z)-2-Penten-1-ol | 91 |
| (Z)-Pent-2-enyl hexanoate | 92 |
| 4-Pentenyl isothiocyanate | 107 |
| (E,Z)-Phytol | 98 |
| (E,Z)-Phytyl acetate | 98 |
| 3-Pinanone | 103 |
| l-Piperitone | 101 |
| Prenyl acetate | 97 |
| Prenyl caproate | 97 |
| Prenyl formate | 97 |
| Prenyl isobutyrate | 97 |
| Safrole | 91 |
| Sodium 4-methoxybenzoyloxyacetate | 105 |
| (E)-Tetradec-2-enal | 93 |
| Thujyl alcohol | 103 |
| 2,6,6-Trimethylcyclohex-2-ene-1,4-dione | 102 |
| (E,Z)-3,7,11-Trimethyldodeca-2,6,10-trienyl acetate | 97 |
| 10-Undecen-2-one | 100 |
| Vanillin 3-(l-menthoxy)propane-1,2-diol acetal | 105 |
| Vanillin propylene glycol acetal | 105 |
| Verbenone | 103 |
| Vetiverol | 103 |
| Vetiveryl acetate | 103 |

# ANNEX 1: SUMMARY OF RECOMMENDATIONS FROM THE 69TH JECFA

## Toxicological recommendations and information on specifications

### 1. Food additives and ingredients evaluated toxicologically or assessed for dietary exposure

| Food additive | Specifications[a] | Acceptable daily intake (ADI) and other toxicological recommendations |
|---|---|---|
| **Asparaginase from *Aspergillus niger* expressed in *A. niger*** | N | ADI "**not specified**"[b] when used in the applications specified and in accordance with good manufacturing practice. |
| **Ethyl lauroyl arginate** | N | **ADI of 0–4 mg/kg bw** for Ethyl-$N^\alpha$-lauroyl-L-arginate HCl based on a NOAEL of 442 mg/kg bw per day in two reproductive toxicity studies and a safety factor of 100.<br>The Committee noted that some of the estimates of high exposure (greater than 95th percentile) exceeded the ADI, but recognized that these estimates were highly conservative and that actual intakes were likely to be within the ADI. |
| **Calcium lignosulfonate (40-65)**<br>The suffix (40-65) reflects the weight-average molecular weight range (40 000–65 000) to distinguish it from other calcium lignosulfonates in commerce | N | **ADI of 0–20 mg/kg bw** based on a NOEL of 2000 mg/kg bw per day from a 90-day toxicity study and a safety factor of 100.<br>The maximum potential dietary exposure to calcium lignosulfonate (40–65) was low and not expected to exceed 7 mg/kg bw per day from use as a carrier of fat-soluble vitamins and carotenoids in food and supplements. |
| **Paprika extract**<br>Since the source material and the manufacturing process differ for paprika preparations used as a spice and as a food colour, the name "paprika extract" was adopted for use as a food colour, leaving the term "paprika oleoresin" for use as a spice. | N,T | The Committee did **not allocate an ADI**. Concern was expressed as to whether the material tested in the 90-day and long-term studies was representative of all commercial production of paprika extract used as food colour. The fact that the material tested contained less than 0.01% capsaicin and the fact that the Committee did not receive adequate data to establish a limit for capsaicin in the specifications for paprika extract added to this concern.<br>New tentative specifications were prepared, pending receipt of additional information on paprika extract used as food colour, including concentrations of capsaicin (to differentiate from materials used as flavours) and additional information about the composition of batches of extract produced by a variety of manufacturers. |
| **Phospholipase C expressed in *Pichia pastoris*** | N | ADI "**not specified**"[b] when used in the applications specified and in accordance with good manufacturing practice. |

| Food additive | Specifications[a] | Acceptable daily intake (ADI) and other toxicological recommendations |
|---|---|---|
| **Phytosterols, phytostanols and their esters** | N | **Group ADI of 0–40 mg/kg bw for phytosterols, phytostanols and their esters, expressed as the sum of phytosterols and phytostanols in their free form**, based on an overall NOAEL of 4200 mg/kg bw per day to which a safety factor of 100 was applied. The overall NOAEL was identified using the combined evidence from several studies of short-term (90 day) toxicity. The Committee considered the margin between this overall NOAEL and the lowest LOAEL from the 90 day toxicity studies of 9000 mg/kg bw per day as adequate for this overall NOAEL to be used as the basis for establishing an ADI. This conclusion is supported by the results of the available studies of reproductive toxicity,<br>Based on available data the Committee concluded that dietary exposure to phytosterols and -stanols would typically be within the ADI. |
| **Polydimethylsiloxane (PDMS)** | R | **Temporary ADI of 0–0.8 mg/kg bw for PDMS,** based on the previous ADI and **applying an additional safety factor of 2.** The previously established ADI of 0–1.5 mg/kg bw was withdrawn. Results of studies to elucidate the mechanism and relevance of ocular toxicity observed in the submitted toxicology studies, as well as data on actual use levels in foods should be provided before the end of 2010.<br>The temporary ADI applies to PDMS that meets the revised specifications prepared. |
| **Steviol glycosides** | R | **ADI of 0–4 mg/kg bw expressed as steviol**, based on a NOEL of 970 mg/kg bw per day from a long-term experimental study with stevioside (383 mg/kg bw per day expressed as steviol) and a safety factor of 100. The results of the new studies presented to the Committee showed no adverse effects of steviol glycosides when taken at doses of about 4 mg/kg bw per day, expressed as steviol, for up to 16 weeks by individuals with type 2 diabetes mellitus and individuals with normal or low-normal blood pressure for 4 weeks.<br><br>Some estimates of high-percentile dietary exposure to steviol glycosides exceeded the ADI, particularly when assuming complete replacement of caloric sweeteners with steviol glycosides. The Committee recognized that these estimates were highly conservative and that actual intakes were likely to be within the ADI. |
| **Sulfites**<br>Dietary exposure assessment | | The main contributors to total dietary exposure to sulfites differ between countries owing to differing patterns of use of sulfites in foods and of consumption of foods to which sulfites may be added. Thus dried fruit, sausages and nonalcoholic beverages were the main contributors of sulfites in some countries, while in other countries these foods are generally produced without the use of sulfites. In countries where wine is regularly consumed, it was one of the main contributors to dietary exposure in adults. Dietary exposure in high regular consumers of wine (97.5[th] percentile) was shown to exceed the ADI for sulfites (0-0.7 mg/kg bw) based either on MLs in Codex GSFA, on MLs in national legislation or on the average concentration determined analytically (about 100 mg/l). |

|  |  | In children and teenagers, a significant contribution to mean exposure to sulfites could come from fruit juices and soft drinks (including cordial), sausages, various forms of processed potatoes, dried fruit and nuts. |
|  |  | Other significant contributions to dietary exposure in the adult population come from dried fruit, sausages and beer. |
|  |  | The Committee provided recommendation on further relevant actions to be considered by countries and the Codex Alimentarius Commission (see Annex 2). |

[a] N: new specifications prepared; R: existing specifications revised; S: existing specifications maintained; T: tentative specifications.

[b] ADI 'not specified' is used to refer to a food substance of very low toxicity which, on the basis of the available data (chemical, biochemical, toxicological and other) and the total dietary intake of the substance arising from its use at the levels necessary to achieve the desired effects and from its acceptable background levels in food, does not, in the opinion of the Committee, represent a hazard to health. For that reason, and for the reasons stated in the individual evaluations, the establishment of an ADI expressed in numerical form is not deemed necessary. An additive meeting this criterion must be used within the bounds of good manufacturing practice, i.e. it should be technologically efficacious and should be used at the lowest level necessary to achieve this effect, it should not conceal food of inferior quality or adulterated food, and it should not create a nutritional imbalance.

## 2. Food additives, including flavouring agents, considered for specifications only

| Food Additive | Specifications[a] |
|---|---|
| Canthaxanthin | R |
| Carob bean gum and carob bean gum (clarified) | R |
| Chlorophyllin, copper complexes sodium and potassium salts | R |
| Carbohydrase from *Aspergillus niger var.* | W |
| Estragole | W |
| Fast Green FCF | R |
| Guar gum and guar gum (clarified) | R |
| Iron oxides | R |
| Isomalt | R |
| Monomagnesium phosphate | N |
| Patent Blue V | R |
| Sunset Yellow FCF | R |
| Trisodium diphosphate | N |

[a] N: New specifications prepared; R: Existing specifications revised; T: tentative specifications; W: Existing specifications withdrawn.

## 3. Flavouring agents

### 3.1. Flavourings evaluated by the Procedure for the Safety Evaluation of Flavouring Agents

#### 3.1.1 Aliphatic, linear α,β-unsaturated aldehydes, acids and related alcohols, acetals and esters

| Flavouring agent | No. | Specifications[a] | Conclusions based on current estimated intake |
|---|---|---|---|
| ***Structural Class I*** | | | |
| (Z)-2-Penten-1-ol | 1793 | N | No safety concern |
| (E)-2-Decen-1-ol | 1794 | N | No safety concern |
| (Z)-Pent-2-enyl hexanoate | 1795 | N | No safety concern |
| (E)-2-Hexenyl octanoate | 1796 | N | No safety concern |
| trans-2-Hexenyl 2-methylbutyrate | 1797 | N | No safety concern |
| Hept-trans-2-en-1-yl acetate | 1798 | N | No safety concern |
| (E,Z)-Hept-2-en-1-yl isovalerate | 1799 | N | No safety concern |
| trans-2-Hexenal glyceryl acetal | 1800 | N | No safety concern |
| trans-2-Hexenal propylene glycol acetal | 1801 | N | No safety concern |
| cis- and trans-1-Methoxy-1-decene | 1802 | N | No safety concern |
| (E)-Tetradec-2-enal | 1803 | N | No safety concern |
| (E)-2-Pentenoic acid | 1804 | N | No safety concern |
| (E)-2-Octenoic acid | 1805 | N | No safety concern |
| Ethyl trans-2-butenoate | 1806 | N | No safety concern |
| Hexyl 2-butenoate | 1807 | N | No safety concern |
| Ethyl trans-2-hexenoate | 1808 | N | No safety concern |
| (E,Z)-Methyl 2-hexenoate | 1809 | N | No safety concern |
| Hexyl trans-2-hexenoate | 1810 | N | No safety concern |
| Methyl trans-2-octenoate | 1811 | N | No safety concern |
| Ethyl trans-2-octenoate | 1812 | N | No safety concern |
| (E,Z)-Methyl 2-nonenoate | 1813 | N | No safety concern |
| Ethyl trans-2-decenoate | 1814 | N | No safety concern |

[a]N: new specifications prepared

#### 3.1.2 Aliphatic branched-chain saturated and unsaturated alcohols, aldehydes, acids, and related esters

| Flavouring agent | No. | Specifications[a] | Conclusions based on current estimated intake |
|---|---|---|---|
| ***Structural class I*** | | | |
| Ethyl (E)-2-methyl-2-pentenoate | 1815 | N | No safety concern |
| 2-Methylbutyl 3-methyl-2-butenoate | 1816 | N | No safety concern |
| (+/-)(E,Z)-5-(2,2-Dimethylcyclopropyl)-3-methyl-2-pentenal | 1817 | N | No safety concern |
| (E,Z)-4-Methylpent-2-enoic acid | 1818 | N | No safety concern |
| (+/-)-4-Ethyloctanal | 1819 | N | No safety concern |
| (E)-Geranyl 2-methylbutyrate | 1820 | N | No safety concern |
| (E)-Geranyl valerate | 1821 | N | No safety concern |
| (E)-Geranyl tiglate | 1822 | N | No safety concern |
| (E)-Citronellyl 2-methylbut-2-enoate | 1823 | N | No safety concern |
| (E)-Ethyl tiglate | 1824 | N | No safety concern |
| (E,Z)-Geranic acid | 1825 | N | No safety concern |
| Prenyl formate | 1826 | N | No safety concern |
| Prenyl acetate | 1827 | N | No safety concern |

| Flavouring agent | No. | Specifications[a] | Conclusions based on current estimated intake |
|---|---|---|---|
| Prenyl isobutyrate | 1828 | N | No safety concern |
| Prenyl caproate | 1829 | N | No safety concern |
| (+/-)-Dihydrofarnesol | 1830 | N | No safety concern |
| (E,Z)-3,7,11-Trimethyldodeca-2,6,10-trienyl acetate | 1831 | N | No safety concern |
| (E,Z)-Phytol | 1832 | N | No safety concern |
| (E,Z)-Phytyl acetate | 1833 | N | No safety concern |
| *Structural class II* | | | |
| Methyl 2-methyl-2-propenoate | 1834 | N | No safety concern |

[a] N: new specifications prepared

### 3.1.3 Aliphatic secondary alcohols, ketones and related esters

| Flavouring agent | No. | Specifications[a] | Conclusions based on current estimated intake |
|---|---|---|---|
| *Structural class I* | | | |
| Isopropenyl acetate | 1835 | N | No safety concern |
| 1-Octen-3-yl acetate | 1836 | N | No safety concern |
| 1-Octen-3-yl butyrate | 1837 | N | No safety concern |
| 6-Methyl-5-hepten-2-yl acetate | 1838 | N | No safety concern |
| 3-(Hydroxymethyl)-2-octanone | 1839 | N | No safety concern |
| (+/-)-[R-(E)]-5-Isopropyl-8-methylnona-6,8-dien-2-one | 1840 | N | No safety concern |
| (+/-)-cis- and trans-4,8-Dimethyl-3,7-nonadien-2-ol | 1841 | N | No safety concern |
| 2,4-Dimethyl-4-nonanol | 1850 | N | No safety concern |
| *Structural class II* | | | |
| (+/-)-1-Hepten-3-ol | 1842 | N | No safety concern |
| (E, Z)-4-Octen-3-one | 1843 | N | No safety concern |
| (E)-2-Nonen-4-one | 1844 | N | No safety concern |
| (E)-5-Nonen-2-one | 1845 | N | No safety concern |
| (Z)-3-Hexenyl 2-oxopropionate | 1846 | N | No safety concern |
| (+/-)-cis- and trans-4,8-Dimethyl-3,7-nonadien-2-yl acetate | 1847 | N | No safety concern |
| (E)-1,5-Octadien-3-one | 1848 | N | No safety concern |
| 10-Undecen-2-one | 1849 | N | No safety concern |
| 8-Nonen-2-one | 1851 | N | No safety concern |

[a] N: new specifications prepared.

### 3.1.4 Substances structurally related to menthol

| Flavouring agent | No. | Specifications[a] | Conclusions based on current estimated intake |
|---|---|---|---|
| *Structural Class I* | | | |
| Menthyl valerate | 1852 | N | No safety concern |
| 2-(l-Menthoxy)ethanol | 1853 | N | No safety concern |
| l-Menthyl acetoacetate | 1854 | N | No safety concern |
| l-Menthyl (R,S)-3-hydroxybutyrate | 1855 | N | No safety concern |
| 8-p-Menthene-1,2-diol | 1860 | N | No safety concern |
| *Structural Class II* | | | |
| l-Piperitone | 1856 | N | No safety concern |
| 2,6,6-Trimethylcyclohex-2-ene-1,4-dione | 1857 | N | No safety concern |
| Menthyl pyrrolidone carboxylate | 1858 | N | No safety concern |
| 3,9-Dimethyl-6-(1-methylethyl)-1,4-dioxaspiro[4.5]decan-2-one | 1859 | N | No safety concern |

| Flavouring agent | No. | Specifications[a] | Conclusions based on current estimated intake |
|---|---|---|---|
| d-2,8-p-Menthadien-1-ol | 1861 | N | No safety concern |

[a]N: new specifications prepared.

### 3.1.5 Monocyclic and bicyclic secondary alcohols, ketones and related esters

| Flavouring agent | No. | Specifications[a] | Conclusions based on current estimated intake |
|---|---|---|---|
| *Structural Class I* | | | |
| Dehydronootkatone | 1862 | N | No safety concern |
| Isobornyl isobutyrate | 1863 | N | No safety concern |
| l-Bornyl acetate | 1864 | N | No safety concern |
| Thujyl alcohol | 1865 | N | No safety concern |
| *Structural class II* | | | |
| Vetiverol | 1866 | N | No safety concern |
| Vetiveryl acetate | 1867 | N | No safety concern |
| 3-Pinanone | 1868 | N | No safety concern |
| Isobornyl 2-methylbutyrate | 1869 | N | No safety concern |
| Verbenone | 1870 | N | No safety concern |

[a]N: new specifications prepared.

### 3.1.6 Aliphatic acyclic primary alcohols with aliphatic linear saturated carboxylic acids

| Flavouring agent | No. | Specifications[a] | Conclusions based on current estimated intake |
|---|---|---|---|
| *Structural class I* | | | |
| Methyl hexanoate | 1871 | N | No safety concern |
| Hexyl heptanoate | 1872 | N | No safety concern |
| Hexyl nonanoate | 1873 | N | No safety concern |
| Hexyl decanoate | 1874 | N | No safety concern |
| Heptyl heptanoate | 1875 | N | No safety concern |
| Dodecyl propionate | 1876 | N | No safety concern |
| Dodecyl butyrate | 1877 | N | No safety concern |

[a]N: new specifications prepared.

### 3.1.7 Hydroxy- and alkoxy- substituted benzyl derivatives

| Flavouring agent | No. | Specifications[a] | Conclusions based on current estimated intake |
|---|---|---|---|
| *Structural class I* | | | |
| 4-Hydroxy-3,5-dimethoxy benzaldehyde | 1878 | N | No safety concern |
| Vanillin 3-(l-menthoxy)propane-1,2-diol acetal | 1879 | N | No safety concern |
| Sodium 4-methoxybenzoyloxyacetate | 1880 | N | No safety concern |
| Vanillin propylene glycol acetal | 1882 | N | No safety concern |
| 4-Methoxybenzoyloxyacetic acid | 1883 | N | No safety concern |
| *Structural class III* | | | |
| Divanillin | 1881 | N | No safety concern |

[a]N: new specifications prepared.

### 3.1.8 Miscellaneous nitrogen-containing substances

| Flavouring agent | No. | Specifications[a] | Conclusions based on current estimated intake |
|---|---|---|---|
| *Structural class II* | | | |
| Methyl isothiocyanate | 1884 | N | No safety concern |
| Ethyl isothiocyanate | 1885 | N | No safety concern |
| Isobutyl isothiocyanate | 1886 | N | No safety concern |
| Isoamyl isothiocyanate | 1887 | N | No safety concern |
| Isopropyl isothiocyanate | 1888 | N | No safety concern |
| 3-Butenyl isothiocyanate | 1889 | N | No safety concern |
| 2-Butyl isothiocyanate | 1890 | N | No safety concern |
| 4-(Methylthio)butyl isothiocyanate | 1892 | N | No safety concern |
| 4-Pentenyl isothiocyanate | 1893 | N | No safety concern |
| 5-Hexenyl isothiocyanate | 1894 | N | No safety concern |
| 5-(Methylthio)pentyl isothiocyanate | 1896 | N | No safety concern |
| 6-(Methylthio)hexyl isothiocyanate | 1897 | N | No safety concern |
| *Structural class III* | | | |
| Amyl isothiocyanate | 1891 | N | No safety concern |
| Hexyl isothiocyanate | 1895 | N | No safety concern |

[a]N: new specifications prepared.

### 3.1.9 Furan-substituted aliphatic hydrocarbons, alcohols, aldehydes, ketones, carboxylic acids and related esters, sulfides, disulfides and ethers

The Committee concluded that the Procedure could not be applied to this group, because of the unresolved toxicological concerns. Studies that would assist in the safety evaluation include investigations of the influence of the nature and position of ring substitution on metabolism and on covalent binding to macromolecules. Depending on the findings, additional studies might include assays related to the mutagenic and carcinogenic potential of representative members of this group.

| Flavouring agent | JECFA No. | Specifications[a] |
|---|---|---|
| *Structural Class II* | | |
| 2-Methylfuran | 1487 | S |
| 2,5-Dimethylfuran | 1488 | S |
| 2-Ethylfuran | 1489 | S |
| 2-Butylfuran | 1490 | S |
| 2-Pentylfuran | 1491 | S |
| 2-Heptylfuran | 1492 | S |
| 2-Decylfuran | 1493 | S |
| 3-Methyl-2-(3-methylbut-2-enyl)-furan | 1494 | S |
| 3-(2-Furyl)acrolein | 1497 | S |
| 3-(5-Methyl-2-furyl)prop-2-enal | 1499 | S |
| 2-Furyl methyl ketone | 1503 | S |
| 2-Acetyl-5-methylfuran | 1504 | S |
| 2-Acetyl-3,5-dimethylfuran | 1505 | S |
| 2-Butyrylfuran | 1507 | S |
| (2-Furyl)-2-propanone | 1508 | S |
| 2-Pentanoylfuran | 1509 | S |
| 1-(2-Furyl)butan-3-one | 1510 | S |
| 4-(2-Furyl)-3-buten-2-one | 1511 | S |
| Ethyl 3-(2-furyl)propanoate | 1513 | S |
| Isobutyl 3-(2-furan)propionate | 1514 | S |
| Isoamyl 3-(2-furan)propionate | 1515 | S |

| Flavouring agent | JECFA No. | Specifications[a] |
|---|---|---|
| Isoamyl 4-(2-furan)butyrate | 1516 | S |
| Phenethyl 2-furoate | 1517 | S |
| Furfuryl methyl ether | 1520 | S |
| Ethyl furfuryl ether | 1521 | S |
| Difurfuryl ether | 1522 | S |
| 2,5-Dimethyl-3-furanthiol acetate | 1523 | S |
| Furfuryl 2-methyl-3-furyl disulfide | 1524 | S |
| 3-[(2-Methyl-3-furyl)thio]-2-butanone | 1525 | S |
| O-Ethyl S-(2-furylmethyl)thiocarbonate | 1526 | S |
| **Structural Class III** | | |
| 2,3-Dimethylbenzofuran | 1495 | S |
| 2,4-Difurfurylfuran | 1496 | S |
| 2-Methyl-3(2-furyl)acrolein | 1498 | S |
| 3-(5-Methyl-2-furyl)-butanal | 1500 | S |
| 2-Furfurylidene-butyraldehyde | 1501 | S |
| 2-Phenyl-3-(2-furyl)prop-2-enal | 1502 | S |
| 3-Acetyl-2,5-dimethylfuran | 1506 | S |
| Pentyl 2-furyl ketone | 1512 | S |
| Propyl 2-furanacrylate | 1518 | S |
| 2,5-Dimethyl-3-oxo-(2H)-fur-4-yl butyrate | 1519 | S |

[a]S: Specifications maintained. The specifications monographs will include a statement that the safety evaluation has not been completed.

### 3.1.10 Alkoxy-substituted allylbenzenes present in foods, essential oils, and used as flavouring agents

The Committee concluded that the data reviewed on the six alkoxy-substituted allylbenzenes provide evidence of toxicity and carcinogenicity to rodents given high doses for several of these substances. A mechanistic understanding of these effects and their implications for human risk have yet to be fully explored, and will have a significant impact on the assessment of health risks from alkoxy-substituted allylbenzenes at the concentrations at which they occur in food.

| Flavouring agent | No. | Specifications[a] |
|---|---|---|
| Apiole | 1787 | N |
| Elemicin | 1788 | N |
| Estragole | 1789 | N |
| Methyl eugenol | 1790 | N |
| Myristicin | 1791 | N |
| Safrole | 1792 | N |

[a]N: new specifications prepared. The specifications monographs will include a statement that the safety evaluation has not been completed.

### 3.2 Re-evaluation of safety of certain flavourings

At the fifty-ninth, sixty-first, sixty-third and sixty-fifth meetings of the Committee, only "anticipated" annual volumes of productions were provided for some flavouring agents and used in the MSDI calculation. These volumes were used for expedience in completing a safety evaluation, but the conclusions of the Committee were made conditional pending the submission of actual poundage data.

Actual production volumes were subsequently submitted for all 143 requested flavouring agents and were evaluated by the Committee. The two flavouring substances requiring a re-evaluation were No. 1414, l-monomenthyl glutarate and No. 1595, 2-isopropyl-N,2,3-trimethylbutyramide.

The Committee concluded that the Procedure could not be applied to 2-isopropyl-N,2,3-trimethylbutyramide, because of evidence of clastogenicity in the presence, but not in the absence, of metabolic activation.

| Flavouring agent | No. | Specifications[a] | Conclusions based on current estimated intake |
|---|---|---|---|
| Ethyl cyclohexanecarboxylate | 963 | S | No safety concern |
| 10-Hydroxymethylene-2-pinene | 986 | S | No safety concern |
| 2,5-Dimethyl-3-furanthiol | 1063 | S | No safety concern |
| Propyl 2-methyl-3-furyl disulfide | 1065 | S | No safety concern |
| Bis(2-methyl-3-furyl) disulfide | 1066 | S | No safety concern |
| Bis(2,5-dimethyl-3-furyl) disulfide | 1067 | S | No safety concern |
| Bis(2-methyl-3-furyl) tetrasulfide | 1068 | S | No safety concern |
| 2,5-Dimethyl-3-furan thioisovalerate | 1070 | S | No safety concern |
| Furfuryl isopropyl sulfide | 1077 | S | No safety concern |
| 2-Methyl-3,5- or 6-(furfurylthio)pyrazine | 1082 | S | No safety concern |
| 3-[(2-Methyl-3-furyl)thio]-4-heptanone | 1085 | S | No safety concern |
| 2,6-Dimethyl-3-[(2-methyl-3-furyl)thio]-4-heptanone | 1086 | S | No safety concern |
| 4-[(2-Methyl-3-furyl)thio]-5-nonanone | 1087 | S | No safety concern |
| 2-Methyl-3-thioacetoxy-4,5-dihydrofuran | 1089 | S | No safety concern |
| 4-Hydroxy-4-methyl-5-hexenoic acid gamma-lactone | 1157 | S | No safety concern |
| (+/-) 3-Methyl-gamma-decalactone | 1158 | S | No safety concern |
| 4-Hydroxy-4-methyl-7-cis-decenoic acid gamma-lactone | 1159 | S | No safety concern |
| Tuberose lactone | 1160 | S | No safety concern |
| Dihydromintlactone | 1161 | S | No safety concern |
| Mintlactone | 1162 | S | No safety concern |
| Dehydromenthofurolactone | 1163 | S | No safety concern |
| (+/-)-(2,6,6-Trimethyl-2-hydroxycyclohexylidene) acetic acid gamma-lactone | 1164 | S | No safety concern |
| 2-(4-Methyl-2-hydroxyphenyl)propionic acid gamma-lactone | 1167 | S | No safety concern |
| 2,4-Hexadien-1-ol | 1174 | S | No safety concern |
| (E,E)-2,4-Hexadienoic acid | 1176 | S | No safety concern |
| (E,E)-2,4-Octadien-1-ol | 1180 | S | No safety concern |
| 2,4-Nonadien-1-ol | 1183 | S | No safety concern |
| (E,Z)-2,6-Nonadien-1-ol acetate | 1188 | S | No safety concern |
| (E,E)-2,4-Decadien-1-ol | 1189 | S | No safety concern |
| Methyl (E)-2-(Z)-4-decadienoate | 1191 | S | No safety concern |
| Ethyl 2,4,7-decatrienoate | 1193 | S | No safety concern |
| (+/-) 2-Methyl-1-butanol | 1199 | S | No safety concern |
| 2-Methyl-2-octenal | 1217 | S | No safety concern |
| 4-Ethyloctanoic acid | 1218 | S | No safety concern |
| 8-Ocimenyl acetate | 1226 | S | No safety concern |
| 3,7,11-Trimethyl-2,6,10-dodecatrienal | 1228 | S | No safety concern |
| 12-Methyltridecanal | 1229 | S | No safety concern |
| 1-Ethoxy-3-methyl-2-butene | 1232 | S | No safety concern |
| 2,2,6-Trimethyl-6-vinyltetrahydropyran | 1236 | S | No safety concern |
| Cycloionone | 1239 | S | No safety concern |
| 2,4-Dimethylanisole | 1245 | S | No safety concern |
| 1,2-Dimethoxybenzene | 1248 | S | No safety concern |
| 4-Propenyl-2,6-dimethoxyphenol | 1265 | S | No safety concern |
| erythro and threo-Mercapto-2-methylbutan-1-ol | 1289 | S | No safety concern |
| (±)2-Mercapto-2-methylpentan-1-ol | 1290 | S | No safety concern |
| 3-Mercapto-2-methylpentanal | 1292 | S | No safety concern |
| 4-Mercapto-4-methyl-2-pentanone | 1293 | S | No safety concern |

| Flavouring agent | No. | Specifications[a] | Conclusions based on current estimated intake |
|---|---|---|---|
| spiro[2,4-Dithia-1-methyl-8-oxabicyclo(3.3.0)octane-3,3'-(1'-oxa-2'-methyl)-cyclopentane] | 1296 | S | No safety concern |
| 2,3,5-Trithiahexane | 1299 | S | No safety concern |
| Diisopropyl trisulfide | 1300 | S | No safety concern |
| 2-(2-Methylpropyl)pyridine | 1311 | S | No safety concern |
| 2-Propionylpyrrole | 1319 | S | No safety concern |
| 2-Propylpyridine | 1322 | S | No safety concern |
| 4-Methylbiphenyl | 1334 | S | No safety concern |
| delta-3-Carene | 1342 | S | No safety concern |
| Farnesene (alpha and beta) | 1343 | S | No safety concern |
| 1-Methyl-1,3-cyclohexadiene | 1344 | S | No safety concern |
| trans-2-Octen-1-yl acetate | 1367 | S | No safety concern |
| trans-2-Octen-1-yl butanoate | 1368 | S | No safety concern |
| cis-2-Nonen-1-ol | 1369 | S | No safety concern |
| (E)-2-Octen-1-ol | 1370 | S | No safety concern |
| (E)-2-Butenoic acid | 1371 | S | No safety concern |
| (E)-2-Decenoic acid | 1372 | S | No safety concern |
| (E)-2-Heptenoic acid | 1373 | S | No safety concern |
| (Z)-2-Hexen-1-ol | 1374 | S | No safety concern |
| trans-2-Hexenyl butyrate | 1375 | S | No safety concern |
| (E)-2-Hexenyl formate | 1376 | S | No safety concern |
| trans-2-Hexenyl isovalerate | 1377 | S | No safety concern |
| trans-2-Hexenyl propionate | 1378 | S | No safety concern |
| trans-2-Hexenyl pentanoate | 1379 | S | No safety concern |
| (E)-2-Nonenoic acid | 1380 | S | No safety concern |
| (E)-2-Hexenyl hexanoate | 1381 | S | No safety concern |
| (Z)-3- & (E)-2-Hexenyl propionate | 1382 | S | No safety concern |
| 2-Undecen-1-ol | 1384 | S | No safety concern |
| Dihydronootkatone | 1407 | S | No safety concern |
| beta-Ionyl acetate | 1409 | S | No safety concern |
| alpha-Isomethylionyl acetate | 1410 | S | No safety concern |
| 3-(l-Menthoxy)-2-methylpropane-1,2-diol | 1411 | S | No safety concern |
| Bornyl butyrate | 1412 | S | No safety concern |
| d,l-Menthol-(±)-propylene glycol carbonate | 1413 | S | No safety concern |
| l-Monomenthyl glutarate | 1414 | S | No safety concern |
| l-Menthyl methyl ether | 1415 | S | No safety concern |
| p-Menthane-3,8-diol | 1416 | S | No safety concern |
| Taurine | 1435 | S | No safety concern |
| L-Arginine | 1438 | S | No safety concern |
| L-Lysine | 1439 | S | No safety concern |
| Tetrahydrofurfuryl cinnamate | 1447 | S | No safety concern |
| (±)-2-(5-Methyl-5-vinyltetrahydrofuran-2-yl)propionaldehyde | 1457 | S | No safety concern |
| Ethyl 2-ethyl-3-phenylpropanoate | 1475 | S | No safety concern |
| 2-Oxo-3-phenylpropionic acid and | 1478 | S | No safety concern |
| Sodium 2-Oxo-3-phenylpropionate | 1479 | S | No safety concern |
| 2-Methyl-3-(1-oxopropoxy)-4H-pyran-4-one | 1483 | S | No safety concern |
| 4-Allylphenol | 1527 | S | No safety concern |
| 2-Methoxy-6-(2-propenyl)phenol | 1528 | S | No safety concern |
| Eugenyl isovalerate | 1532 | S | No safety concern |

| Flavouring agent | No. | Specifications[a] | Conclusions based on current estimated intake |
|---|---|---|---|
| cis-3-Hexenyl anthranilate | 1538 | S | No safety concern |
| Citronellyl anthranilate | 1539 | S | No safety concern |
| Ethyl N-methylanthranilate | 1546 | S | No safety concern |
| Ethyl N-ethylanthranilate | 1547 | S | No safety concern |
| Isobutyl N-methylanthranilate | 1548 | S | No safety concern |
| Methyl N-formylanthranilate | 1549 | S | No safety concern |
| Methyl N-acetylanthranilate | 1550 | S | No safety concern |
| Methyl N,N-dimethylanthranilate | 1551 | S | No safety concern |
| N-Benzoylanthranilic acid | 1552 | S | No safety concern |
| Trimethyloxazole | 1553 | S | No safety concern |
| 2,5-Dimethyl-4-ethyloxazole | 1554 | S | No safety concern |
| 2-Ethyl-4,5-dimethyloxazole | 1555 | S | No safety concern |
| 2-Isobutyl-4,5-dimethyloxazole | 1556 | S | No safety concern |
| 2-Methyl-4,5-benzo-oxazole | 1557 | S | No safety concern |
| 2,4-Dimethyl-3-oxazoline | 1558 | S | No safety concern |
| Butyl isothiocyanate | 1561 | S | No safety concern |
| Benzyl isothiocyanate | 1562 | S | No safety concern |
| Phenethyl isothiocyanate | 1563 | S | No safety concern |
| 4,5-Dimethyl-2-propyloxazole | 1569 | S | No safety concern |
| 4,5-Epoxy-(E)-2-decenal | 1570 | S | No safety concern |
| beta-Ionone epoxide | 1571 | S | No safety concern |
| Epoxyoxophorone | 1573 | S | No safety concern |
| Ethylamine | 1579 | S | No safety concern |
| Propylamine | 1580 | S | No safety concern |
| Isopropylamine | 1581 | S | No safety concern |
| Isobutylamine | 1583 | S | No safety concern |
| sec-Butylamine | 1584 | S | No safety concern |
| Pentylamine | 1585 | S | No safety concern |
| 2-Methylbutylamine | 1586 | S | No safety concern |
| Hexylamine | 1588 | S | No safety concern |
| 2-(4-Hydroxyphenyl)ethylamine | 1590 | S | No safety concern |
| 1-Amino-2-propanol | 1591 | S | No safety concern |
| Butyramide | 1593 | S | No safety concern |
| 1,6-Hexalactam | 1594 | S | No safety concern |
| 2-Isopropyl-N,2,3-trimethylbutyramidee | 1595 | S | **Further information is needed** |
| N-Ethyl (E)-2,(Z)-6-nonadienamide | 1596 | S | No safety concern |
| N-Cyclopropyl (E)-2,(Z)-6-nonadienamide | 1597 | S | No safety concern |
| N-Isobutyl (E,E)-2,4-decadienamide | 1598 | S | No safety concern |
| (±)-N,N-Dimethyl menthyl succinamide | 1602 | S | No safety concern |
| 1-Pyrroline | 1603 | S | No safety concern |
| 2-Acetyl-1-pyrroline | 1604 | S | No safety concern |
| 2-Propionylpyrroline | 1605 | S | No safety concern |
| Isopentylidene isopentylamine | 1606 | S | No safety concern |
| 2-Methylpiperidine | 1608 | S | No safety concern |
| Triethylamine | 1611 | S | No safety concern |
| Tripropylamine | 1612 | S | No safety concern |
| N,N-Dimethylphenethylamine | 1613 | S | No safety concern |
| Trimethylamine oxide | 1614 | S | No safety concern |
| Piperazine | 1615 | S | No safety concern |

# ANNEX 2: RECOMMENDATIONS AND FURTHER INFORMATION REQUIRED

**Paprika extract**

Data on the composition and capsaicin content of batches of paprika extract for use as a colour produced by a variety of manufacturers. Information as to whether the material used in the toxicological tests submitted was representative of all the products in commerce. If not, additional toxicological data on representative material would be needed for the evaluation of paprika extract for use as a colour.

The Committee recommended that the specifications for paprika oleoresin be revised at a future meeting in order to allow the differentiation of paprika extract used as a colour from paprika oleoresin used as a flavour.

**Polydimethylsiloxane**

Results of studies to elucidate the mechanism and relevance of the ocular toxicity observed in the experimental studies and data on actual use levels in foods should be provided before the end of 2010.

**Sulfites – dietary exposure assessment and maximum levels (MLs) in foods**

Countries that have not yet done so could consider collecting data on the current use of sulfites in food and beverages available on their markets and investigating whether dietary exposure in some subpopulations exceeds the ADI. On the basis of this investigation, individual countries and the food industry could consider the possibility of taking one or more of the following measures to reduce dietary exposure to sulfites so that the ADI is not exceeded in the population:

(1) align national legislation with Codex MLs where these are lower;
(2) take action to effectively enforce national MLs;
(3) encourage research on alternative methods of preservation, particularly on applications in which the use of sulfites is responsible for a significant contribution;
(4) take action so that the use of sulfites is reduced in foods where safe alternative solutions are available.

Codex Alimentarius Commission codes of practices for certain groups of food commodities, such as fruit juice, dried fruit and processed meat, could be amended to include suggestions to help countries and the food industry in the implementation of a reduction of the use of sulfites in food.

**Furan-substituted aliphatic hydrocarbons, alcohols, aldehydes, ketones, carboxylic acids and related esters, sulfides, disulfides and ethers (JECFA Nos, Structural Class II: 1487, 1488, 1489, 1490, 1491, 1492, 1493, 1494, 1497, 1499, 1503, 1504, 1505, 1507, 1508, 1509, 1510, 1511, 1513, 1514, 1515, 1516, 1517, 1520, 1521, 1522, 1523, 1524, 1525, 1526; Structural Class III: 1495, 1496, 1498, 1500, 1501, 1502, 1506, 1512, 1518, 1519)**

The Committee concluded that the Procedure could not be applied to this group of flavouring agents, because of the unresolved toxicological concerns. Studies that would assist in the safety evaluation include investigations of the influence of the nature and position of ring substitution on metabolism and on covalent binding to macromolecules. Depending on the findings, additional studies might include assays related to the mutagenic and carcinogenic potential of representative members of this group of flavours.

**Alkoxy-substituted allylbenzenes present in foods, essential oils, and used as flavouring agents (Apiole JECFA No. 1787, Elemicin No. 1788, Estragole No. 1789, Methyl eugenol No. 1790, Myristicin No 1791, Safrole No 1792)**

There is evidence of toxicity and carcinogenicity to rodents given high doses for several of these substances. A mechanistic understanding of these effects and their implications for human risk have yet to be fully explored, and will have a significant impact on the assessment of health risks from alkoxy-substituted allylbenzenes at the concentrations at which they occur in food. Further research is needed to assess the potential risk to human health from low-level dietary exposure to alkoxy-substituted allylbenzenes present in foods and essential oils and used as flavouring agents.

**2-isopropyl-N,2,3-trimethylbutyramide (JECFA No. 1595)**

The Committee concluded that the Procedure could not be applied to 2-isopropyl-N,2,3-trimethylbutyramide, because of because of evidence of clastogenicity in the presence, but not in the absence, of metabolic activation. Information that would assist in resolving the concerns would include data on the potential of this compound to form reactive metabolites and on whether clastogenicity is also expressed in vivo, as well as additional information on the effects found in the kidney (tubular nephrosis, tubular dilatation with granular casts and hyaline droplet formation) at relatively low doses.

# CORRIGENDA

**COMPENDIUM OF FOOD ADDITIVE SPECIFICATIONS**
**FAO FOOD AND NUTRITION PAPER 52, Addendum 9, ROME, 2001.**

Page 129, Flavouring agent **2-Ethyl-6-methyl pyrazine** (JECFA No. 769): The entry on Assay minium % is corrected to exclude the presence of the 2,3-isomer as follows: 95 (sum of 2,5- and 2,6-isomers).

**COMPENDIUM OF FOOD ADDITIVE SPECIFICATIONS**
**FAO FOOD AND NUTRITION PAPER 52, Addendum 11, ROME, 2003.**

Page 120, the name of flavouring agent with JECFA No. 1290 is corrected to **(+/-)2-Mercapto-2-methylpentan-1-ol**.

**COMPENDIUM OF FOOD ADDITIVE SPECIFICATIONS**
**FAO FOOD AND NUTRITION PAPER 52, Addendum 12, ROME, 2004.**

Page 89, Flavouring agent **DL-(3-Amino-3-carboxypropyl)dimethylsulfonium chloride** (JECFA No. 1427: the Chemical Abstract Services number is corrected to 3493-12-7, to reflect the DL-form of the substance, and the missing letter l in sulfonium is added.

**COMPENDIUM OF FOOD ADDITIVE SPECIFICATIONS**
**FAO FOOD AND NUTRITION PAPER 52, Addendum 13, ROME, 2005.**

Page 59, Flavouring agent **2,5-Dimethyl-3-oxo-(2H)-fur-4-yl butyrate** (JECFA No. 1519): the entry on Secondary Components is modified to indicate the concentration ranges as follows: SC: 1-3% 4-Hydroxy-2,5-dimethyl-3(2H)-furanone and 1-3% Butyric acid

Page 59, Flavouring agent **Furfuryl 2-methyl-3-furyl disulfide** (JECFA No. 1524): the entry on Secondary Components is modified to indicate the concentration range as follows: SC: 6-7% Di-(2-methyl-3-furyl) disulfide.

**FAO TECHNICAL PAPERS**

FAO JECFA MONOGRAPHS

1 Combined compendium of food additive specifications – JECFA specifications monographs from 1st to 65th meeting. (E)
Vol. 1 Food additives A – D
Vol. 2 Food additives E – O
Vol. 3 Food additives P – Z
Vol. 4 Analytical methods, test procedures and laboratory solutions

2 Residue evaluation of certain veterinary drugs Joint FAO/WHO Expert Committee on Food Additives 66th meeting 2006 (E)

3 Compendium of food additive specifications - Joint FAO/WHO Expert Committee on Food Additives 67th meeting 2006 (E)

4 Compendium of food additive specifications - Joint FAO/WHO Expert Committee on Food Additives 68th meeting 2006 (E)

Availability: 2008

| | | | | |
|---|---|---|---|---|
| Ar | – Arabic | | Multil | – Multilingual |
| C | – Chinese | | * | Out of print |
| E | – English | | ** | In preparation |
| F | – French | | | |
| P | – Portuguese | | | |
| S | – Spanish | | | |

*The FAO Technical Papers are available through the authorized FAO Sales Agents or directly from Sales and Marketing Group, FAO, Viale delle Terme di Caracalla, 00153 Rome, Italy.*